A U R O R A

DOVER MODERN MATH ORIGINALS

Dover Publications is pleased to announce the publication of the first volumes in our new Aurora Series of original books in mathematics. In this series we plan to make available exciting new and original works in the same kind of well-produced and affordable editions for which Dover has always been known.

Aurora titles currently in the series or in preparation are:

Optimization in Function Spaces by Amol Sasane. 978-0-486-78945-3

The Theory and Practice of Conformal Geometry by Steven G. Krantz. 978-0-486-79344-3

[illegible] Theorems by Albert [illegible] 978-0-486-[illegible]

Elementary Point-Set Topology: A Transition to Advanced Mathematics by André L. Yandl and Adam Bowers. 978-0-486-80349-4

Additional volumes will be announced periodically.

The Dover Aurora Advisory Board:

[illegible]

AURORA

DOVER MODERN MATH ORIGINALS

Dover Publications is pleased to announce the publication of the first volumes in our new Aurora Series of original books in mathematics. In this series we plan to make available exciting new and original works in the same kind of well-produced and affordable editions for which Dover has always been known.

Aurora titles currently in the process of publication are:

Optimization in Function Spaces by Amol Sasane. (978-0-486-78945-3)

The Theory and Practice of Conformal Geometry by Steven G. Krantz. (978-0-486-79344-3)

Numbers: Histories, Mysteries, Theories by Albrecht Beutelspacher. (978-0-486-80348-7)

Elementary Point-Set Topology: A Transition to Advanced Mathematics by André L. Yandl and Adam Bowers. (978-0-486-80349-4)

Additional volumes will be announced periodically.

The Theory and Practice of Conformal Geometry

STEVEN G. KRANTZ
Washington University in St. Louis

DOVER PUBLICATIONS, INC., Mineola, New York

Bibliographical Note

The Theory and Practice of Conformal Geometry is a new work,
first published by Dover Publications, Inc., in 2016.

International Standard Book Number

ISBN-13: 978-0-486-79344-3
ISBN-10: 0-486-79344-3

Manufactured in the United States by RR Donnelley
79344301 2016
www.doverpublications.com

In homage to Lars Ahlfors.

Contents

Preface

The idea of conformal geometry goes back at least to the Riemann Mapping theorem (RMT), which was first conceived by Bernhard Riemann in his doctoral thesis in 1851. It is well known that Riemann's original proof of this seminal result was flawed. A fully rigorous proof did not appear until at least fifty years later. Nonetheless, many mathematicians refer to the RMT as the "greatest theorem of the nineteenth century."

The standard proof of the RMT learned by most students today makes decisive use of the idea of normal family. Certainly Montel's theorem about normal families is one of the cornerstones of modern function theory.

The interaction of conformal mappings with normal families is of interest to us. Normal families have certain natural invariance properties under conformal mappings, and the two sets of ideas work symbiotically to produce beautiful mathematical theories. In particular, the theory of automorphism groups of domains (and, more generally, of automorphism groups of Riemann surfaces) is generally built on normal families arguments.

A unifying idea in this discussion must be invariant metrics. Of course the granddaddy of all invariant metrics is the Poincaré metric on the disc. But today we have the Bergman metric, the Kobayashi/Royden metric, the Carathéodory metric, and many others. These are decisive tools for understanding both normal families and conformal mappings. We intend to develop this theme in the present book.

A second overarching theme is the Schwarz lemma. It is hard to beat the Schwarz lemma for an uncanny combination of simplicity and profound influence. Thanks to the work of Lars Ahlfors in 1939, we now realize that the Schwarz lemma makes a deep statement about Riemannian geometry. So it fits very naturally with the story told in the present book.

More than sixty years ago Zeev Nehari wrote an elegant book on conformal mapping. That book will serve as a foundation for our treatments. Classical topics such as the Schwarzian derivative, the Schwarz-Christoffel mapping, and Schwarz reflection are nicely treated by Nehari. But there are a host of new developments since that time, and we plan to build on the foundation that Nehari laid.

This author has been studying these ideas for forty years, and has helped to develop some of the modern theory. He takes the reader with a basic foundation in complex variable theory to the forefront of some of the modern developments in the subject. Along the way, the reader will be exposed to some beautiful function theory and also to some of the rudiments of geometry and analysis that make this subject so vibrant and lively.

We introduce in this text a few didactic tools to make the reading stimulating and engaging for students:

(1) Each chapter begins with a *Prologue*, introducing students to the key ideas which will unfold in the text that follows.

(2) Each section begins with a *Capsule*, giving a quick preview of that unit of material.

(3) Each key theorem or proposition is preceded by a *Prelude*, putting the result in context and providing motivation.

(4) At key junctures, we include an *Exercise for the Reader:*, to encourage the neophyte to pick up a pencil, do some calculations, and get involved with the material.

We hope that these devices will break up the usual dry exposition of a research monograph and make this text more like an invitation to the subject.

It is a pleasure to thank my editors Don Albers and John Grafton for their constant encouragement and guidance in this project.

St. Louis, Missouri — Steven G. Krantz

The Theory and Practice of Conformal Geometry

Chapter 1

The Riemann Mapping Theorem

Prologue: There is hardly a more profound theorem from nineteenth century complex analysis than the Riemann Mapping theorem. Even to conceive of such a theorem is virtually miraculous. Although Riemann's original proof was flawed, it pointed in the right direction. Certainly a great deal of modern complex function theory has been inspired by the Riemann Mapping theorem (RMT).

Throughout this book, we shall use the term *domain* to mean a connected, open set. While the Riemann Mapping theorem gives us a complex-analytic classification of *simply connected* planar domains, a theory (in fact several theories) has developed for multiply connected domains. This includes the Ahlfors map, the canonical representation, and the uniformization theorem. We treat all of these in the present chapter. Although we do not treat the topic here, Riemann surface theory is an outgrowth of the study of conformal mappings.

Perhaps the most important modern concept in this circle of ideas is Teichmüller theory, which creates a moduli space for Riemann surfaces. It is beyond the scope of the present book, but it provides a pointer for further reading.

1.0 Introduction

Capsule: It is natural to think of the Riemann Mapping theorem in the context of simply connected domains. However, from the point of view of analysis, it is more convenient to have a different formulation of the topological condition. In this section we introduce the notion of *holomorphic simple connectivity:* A domain U is *holomorphically simply connected* if any holomorphic function on U has a holomorphic antiderivative.

It is easy to verify that any topologically simply connected domain is holomorphically simply connected. So we certainly suffer no loss of generality to use this substitute idea. It also streamlines our treatment.

In thinking about the topology of the plane, it is natural to ask which planar open sets are homeomorphic to the open unit disc. The startling answer is that the Riemann Mapping theorem tells us that *any* connected, simply connected open set (except the plane) is not only homeomorphic to the disc but conformally equivalent to the disc. One can verify separately, by hand, that the entire plane is also homeomorphic to the disc (but certainly not conformally equivalent).

Riemann's astonishing theorem has many different proofs, and we shall consider some of them here. Some of the proofs are "existence proofs," and some constructive. Some are geometric and some are analytic. The book [BIS] covers ideas connected to the Riemann Mapping theorem comprehensively.

We end this section with a formal enunciation of the Riemann mapping theorem:

> **Theorem (RMT):** Let $U \subseteq \mathbb{C}$ be any simply connected domain that is not conformal to the entire plane. Then U is conformally equivalent to the unit disc.

1.1 The Proof of the Analytic Form of the Riemann Mapping Theorem

Capsule: We actually prove the Riemann Mapping theorem in an appendix. In this section, we set up the proof. We develop the idea of holomorphic simple connectivity, and establish some of its properties. We discuss some of the significance of this important result.

Classical arguments, which may be found in any complex analysis text (see for example [GRK1]), show that topological simple connectivity implies an analytic form of simple connectivity that we now define.

Definition 1.1 A connected open set $U \subseteq \mathbb{C}$ is *holomorphically simply connected* if, for each holomorphic function $f : U \to \mathbb{C}$, there is a holomorphic antiderivative F—that is, a function satisfying $F'(z) = f(z)$ on U.

Example 1.2 Certainly open discs and open rectangles are holomorphically simply connected. One constructs F from f with a simple line integral.

Notice that, on a holomorphically simply connected domain, a nonvanishing, holomorphic function f will have a logarithm—for we can just take an antiderivative of f'/f. Then it follows that such an f will have a square root.

Let $D \subseteq \mathbb{C}$ be the unit disc. Let U be a holomorphically simply connected open set in $\mathbb{C}$ that is not equal to all of $\mathbb{C}$. Fix a point $P \in U$ and set

$$\mathcal{F} = \{f : f \text{ is holomorphic on } U, f : U \to D, \\ f \text{ is one-to-one}, f(P) = 0\}.$$

We shall prove the following three assertions:

(1) $\mathcal{F}$ is nonempty.

(2) There is a function $f_0 \in \mathcal{F}$ such that

$$|f_0'(P)| = \sup_{h \in \mathcal{F}} |h'(P)|.$$

(3) If g is any element of $\mathcal{F}$ such that $|g'(P)| = \sup_{h \in \mathcal{F}} |h'(P)|$, then g maps U *onto* the unit disc D.

These three assertions taken together imply the Riemann Mapping theorem (see the discussion below). The proof of assertion **(1)** is by direct construction. Statement **(2)** is proved with a normal families argument. Statement **(3)** is the least obvious and will require some work: If the conclusion of **(3)** is assumed to be false, then we are able to construct an element $\widehat{g} \in \mathcal{F}$ such that $|\widehat{g}'(P)| > |g'(P)|$.

The proof of the Riemann Mapping theorem (as we know it today, not developed by Riemann but rather by Carathéodory and Koebe and others) is quite standard and can be found in most any text. For completeness we include it in an appendix to this section.

Riemann's original proof of his theorem solved a somewhat different extremal problem related to the Dirichlet problem. His argument was flawed because he, in fact, *assumed* that the extremal problem had a solution. He did not prove it. The proof was rescued some years later by refining the Dirichlet principle used by Riemann. Our modern proof sidesteps those difficulties, and uses Montel's theorem to show that the extremal problem has a solution.

There are a number of other approaches to proving the Riemann Mapping theorem. We shall begin the considerations of this chapter by presenting a quite modern proof based on ideas of Thurston. In fact, Thurston introduced the profound new idea of *circle packing*, and Rodin and Sullivan found a way to prove the Riemann Mapping theorem using circle packing. We present their proof here.

A proof that relies on the solution of the Dirichlet problem follows next. This is quite similar in spirit to Riemann's original approach to the theorem. Then we turn to Ahlfors's generalization of Riemann's theorem.

APPENDIX: Traditional Proof of the Riemann Mapping Theorem

Proof of (1): If U is bounded, then this assertion is easy: If we simply let $a = 1/(2\sup\{|z| : z \in U\})$ and $b = -aP$, then the function $f(z) = az + b$ is in $\mathcal{F}$. If U is unbounded, we must work a bit harder. Since $U \neq \mathbb{C}$, there is a point $Q \notin U$. The function $\phi(z) = z - Q$ is nonvanishing on U, and U is holomorphically simply connected. Therefore there exists a holomorphic function h such that $h^2 = \phi$. Notice that h must be one-to-one since ϕ is. Also there cannot be two distinct points $z_1, z_2 \in U$ such that $h(z_1) = -h(z_2)$ [otherwise $\phi(z_1) = \phi(z_2)$]. Now h is a nonconstant holomorphic function; hence an open mapping. Thus the image of h contains a disc $D(b, r)$. But then the image of h must be disjoint from the disc $D(-b, r)$. We may therefore define the holomorphic function

$$f(z) = \frac{r}{2(h(z) + b)}.$$

Since $|h(z) - (-b)| \geq r$ for $z \in U$, it follows that f maps U to D. Since h is one-to-one, so is f. Composing f with a suitable automorphism of D (a Möbius transformation), we obtain a function that is not only one-to-one and holomorphic with image in the disc, but also maps P to 0. Thus $f \in \mathcal{F}$. □

Proof of (2): Of course we may select a sequence of elements $f_j \in \mathcal{F}$ so that $\lim_{j\to\infty} |f_j'(P)| = \sup_{h\in\mathcal{F}} |h'(P)|$. We want to claim that $\{f_j\}$ is a normal family. But of course each $|f_j|$ is bounded by 1, so Montel's theorem applies.

We may derive the desired conclusion once it has been established that the limit derivative-maximizing function is itself one-to-one. Suppose that the $f_j \in \mathcal{F}$ converge normally to a function f_0, with

$$|f_0'(P)| = \sup_{h\in\mathcal{F}} |h'(P)|.$$

We want to show that f_0 is one-to-one into D. The argument principle, specifically Hurwitz's theorem, will now yield this conclusion:

Fix a point $b \in U$. Consider the holomorphic functions $g_j(z) \equiv f_j(z) - f_j(b)$ on the open set $U \setminus \{b\}$. Each f_j is one-to-one; hence each g_j is nowhere vanishing on $U \setminus \{b\}$. Hurwitz's theorem guarantees that either the limit function $f_0(z) - f_0(b)$ is identically zero or is nowhere vanishing. But, for a function $h \in \mathcal{F}$, it must hold that $h'(P) \neq 0$ because if $h'(P)$ were equal to zero, then that h would not be one-to-one. Since $\mathcal{F}$ is nonempty, it follows that $\sup_{h\in\mathcal{F}} |h'(P)| > 0$. Thus the function f_0, which satisfies $|f_0'(P)| = \sup_{h\in\mathcal{F}} |h'(P)|$, cannot have $f_0'(P) = 0$ and f_0 cannot be constant. The only possible conclusion is that $f_0(z) - f_0(b)$ is nowhere zero on $U \setminus \{b\}$. Since this statement holds for each $b \in U$, we conclude that f_0 is one-to-one. □

Proof of (3): Let $g \in \mathcal{F}$ and suppose that there is a point $R \in D$ such that the image of g does not contain R. Set

$$\phi(z) = \frac{g(z) - R}{1 - g(z)\overline{R}}.$$

Here we have composed g with a transformation that preserves the disc and is one-to-one. Note that ϕ is nonvanishing.

The holomorphic simple connectivity of U guarantees the existence of a holomorphic function $\psi : U \to \mathbb{C}$ such that $\psi^2 = \phi$. Now ψ is still one-to-one and has range contained in the unit disc. However, it cannot be in $\mathcal{F}$ since it is nonvanishing. We repair this by composing with another Möbius transformation. Define

$$\rho(z) = \frac{\psi(z) - \psi(P)}{1 - \psi(z)\overline{\psi(P)}}.$$

Then $\rho(P) = 0$, ρ maps U into the disc, and ρ is one-to-one. Therefore $\rho \in \mathcal{F}$. Now we will calculate the derivative of ρ at P and show that it is actually larger in modulus than the derivative of g at P.

We have

$$\begin{aligned}\rho'(P) &= \frac{(1 - |\psi(P)|^2) \cdot \psi'(P) - (\psi(P) - \psi(P))(-\psi'(P)\overline{\psi(P)})}{(1 - |\psi(P)|^2)^2} \\ &= \frac{1}{1 - |\psi(P)|^2} \cdot \psi'(P).\end{aligned}$$

Also,

$$\begin{aligned}2\psi(P) \cdot \psi'(P) &= \phi'(P) \\ &= \frac{(1 - g(P)\overline{R})g'(P) - (g(P) - R) \cdot (-g'(P)\overline{R})}{(1 - g(P)\overline{R})^2}.\end{aligned}$$

But $g(P) = 0$; hence

$$2\psi(P) \cdot \psi'(P) = (1 - |R|^2)g'(P).$$

We conclude that

$$\begin{aligned}\rho'(P) &= \frac{1}{1 - |\psi(P)|^2} \cdot \frac{1 - |R|^2}{2\psi(P)} g'(P) \\ &= \frac{1}{1 - |\phi(P)|} \cdot \frac{1 - |R|^2}{2\psi(P)} g'(P) \\ &= \frac{1}{1 - |R|} \cdot \frac{1 - |R|^2}{2\psi(P)} g'(P) \\ &= \frac{1 + |R|}{2\psi(P)} g'(P).\end{aligned}$$

However $1 + |R| > 1$ (since $R \neq 0$) and $|\psi(P)| = \sqrt{|R|}$. It follows, since $(1 + |R|)/(2\sqrt{|R|}) > 1$, that

$$|\rho'(P)| > |g'(P)|.$$

Thus, if the mapping g of statement **(3)** at the beginning of the section were not onto, then it could not have property **(2)**, of maximizing the absolute value of the derivative at P. □

We have completed the proofs of each of the three assertions and hence of the analytic form of the Riemann Mapping theorem. For completeness, we again enunciate the result.

Prelude: This is the celebrated Riemann Mapping theorem (RMT). Our three steps converge now rather nicely to the proof of this result.

THEOREM 1.3 *Let U be a planar domain which is not the entire plane and which is analytically simply connected. Then there is a conformal map $\varphi : U \to D$ from U to the unit disc.*

Certainly statements **(2)** and **(3)** taken together give us such a mapping which is both one-to-one and onto.

The proof of statement **(3)** may have seemed unmotivated. Let us have another look at it. Let

$$\mu(z) = \frac{z - R}{1 - \overline{R}z},$$

$$\tau(z) = \frac{z - \psi(P)}{1 - \overline{\psi(P)}z},$$

and

$$S(z) = z^2.$$

Then, by our construction, with $h = \mu^{-1} \circ S \circ \tau^{-1}$:

$$\begin{aligned} g &= \mu^{-1} \circ S \circ \tau^{-1} \circ \rho \\ &\equiv h \circ \rho. \end{aligned}$$

Now the chain rule tells us that

$$|g'(P)| = |h'(0)| \cdot |\rho'(P)|.$$

Since $h(0) = 0$, and since h is not a conformal equivalence of the disc to itself, the Schwarz lemma tells us that $|h'(0)|$ must be less than 1. But this says that $|g'(P)| < |\rho'(P)|$, giving us the required contradiction.

1.2 A New Proof of the RMT

Capsule: In a lecture given in the 1980s, W. Thurston proposed the idea that conformality can be understood in terms of circle packing.

This idea captured the imagination of a number of mathematicians, and a new subject was quickly born.

B. Rodin and D. Sullivan were fairly quickly able to create a circle-packing proof of the Riemann Mapping theorem. This result really put Thurston's idea on the map, and also placed into focus some of the key techniques that would be widely used in this new study.

In this section, we explain the proof of Rodin and Sullivan.

There are many proofs of the Riemann Mapping theorem. Some of them are quite analytical—see, for instance, [BUR, pp. 293 ff.] Others are more geometric. We present one of those here.

In fact, this proof stems from a lecture given by W. P. Thurston in 1985. The basic idea is that we can internally approximate a simply connected planar domain with a circle packing, and then produce a corresponding packing in the unit disc. The natural mapping of the nerves of the packings then gives an approximate conformal mapping. In the limit, the Riemann mapping is produced.

Circle packing has given rise to an entire new approach to complex function theory. The elegant book [STE] is an homage to these new developments. B. Rodin and D. Sullivan [ROS] determined how to make Thurston's ideas concrete, and how to provide a rigorous circle-packing proof of the Riemann Mapping theorem. That is the proof that we present here.

1.2.1 Some Definitions

Let U be a planar domain. A *circle packing* in U is a collection of closed discs contained in U and having disjoint interiors. See Figures 1.1, 1.2, and 1.3 for a variety of circle packings.

The *nerve* of the circle packing is the embedded 1-complex whose vertices are the centers of the discs and whose edges are the geodesic segments joining the centers of tangent discs and passing through the point of tangency. We shall restrict our attention in this discussion to a circle packing whose nerve is the 1-skeleton of a triangulation of some open, connected subset in the plane or the Riemann sphere. In particular, a circle packing of the sphere is one whose associated triangulation is a triangulation of the sphere. It is easy to see that, in a circle packing of the unit disc D, the carrier of the associated triangulation is a proper submanifold of the unit disc.

A finite sequence of circles from a circle packing is called a *chain* if each circle except the last is tangent to its successor. The chain is a *cycle* if the first and last circles are tangent.

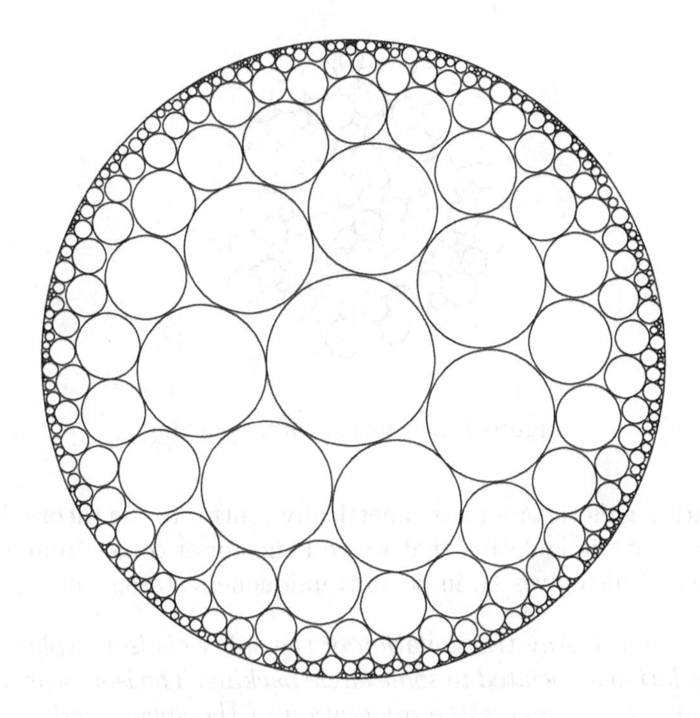

Figure 1.1: A circle packing.

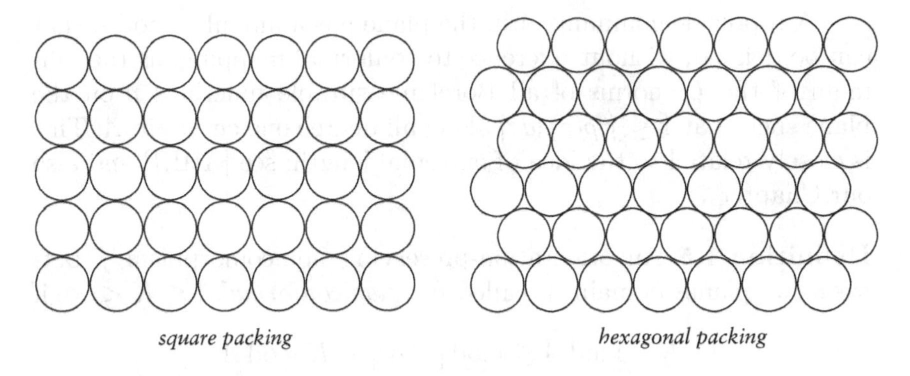

Figure 1.2: A square and a hexagonal circle packing.

Now let c be a circle in a circle packing. The *flower* centered at c is the closed set consisting of c and its interior, together with all circles tangent to c and their interiors, and also the interiors of all the triangular interstices formed by these circles.

We shall not prove the next result, but instead refer the reader to [AND1], [AND2], [THU, Chapter 13], and [ROS, Appendix 2].

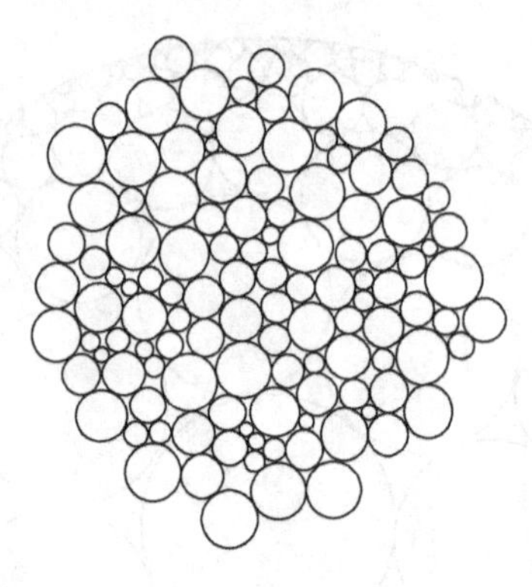

Figure 1.3: Another circle packing.

Prelude: This theorem is geometrically central to the theory. It tells us that any triangulation that we will encounter comes from a circle packing. It also gives an important uniqueness statement.

THEOREM 1.4 *Any triangulation of the sphere is isomorphic to the triangulation associated to some circle packing. The isomorphism can be required to preserve the orientation of the sphere and then this circle packing is unique up to Möbius transformations.*

A topological annulus A in the plane has a modulus $\operatorname{mod} A$ that can be defined, without reference to conformal mapping, as the infimum of the L^2 norms of all Borel measurable functions ρ on the plane such that $1 \le \int \rho(z)\,|dz|$ along all degree one curves in A. This is closely related to the idea of extremal length; see [AHL1] and also our Chapter 7.

Definition 1.5 An orientation-preserving homeomorphism f between two planar domains is called K*–quasiconformal*, $1 \le K < \infty$, if

$$K^{-1} \operatorname{mod} A \le \operatorname{mod} [f(A)] \le K \operatorname{mod} A$$

for every annulus A in the domain of f.

Intuitively, a K-quasiconformal mapping does not distort the picture by more than a factor of K. Some useful facts about quasiconformal mappings are these:

(1) K-quasiconformality is a local property (see [AHL3, page 22]).

(2) A 1-quasiconformal map is conformal and conversely.

Further, we need the fact that a simplicial homeomorphism is K-quasiconformal for K depending only on the shapes of the triangles involved.

1.2.2 A Sketch of Thurston's Idea

Let U be a simply connected, proper subdomain of the plane. Now efficiently fill U with small circles from a regular, hexagonal circle packing. Surround these circles by a Jordan curve. Use Andreeev's theorem to produce a combinatorially equivalent packing of the unit disc, with the unit circle corresponding to the Jordan curve. What we hope is that the correspondence between the circles of the two packings will approximate the Riemann mapping.

1.2.3 Details of the Proof

LEMMA 1.6 (THE RING LEMMA) *There is a constant r depending only on k such that, if k circles surround the unit disc (i.e., they form a cycle externally tangent to the unit disc), then each circle has radius at least r. See Figure 1.4.*

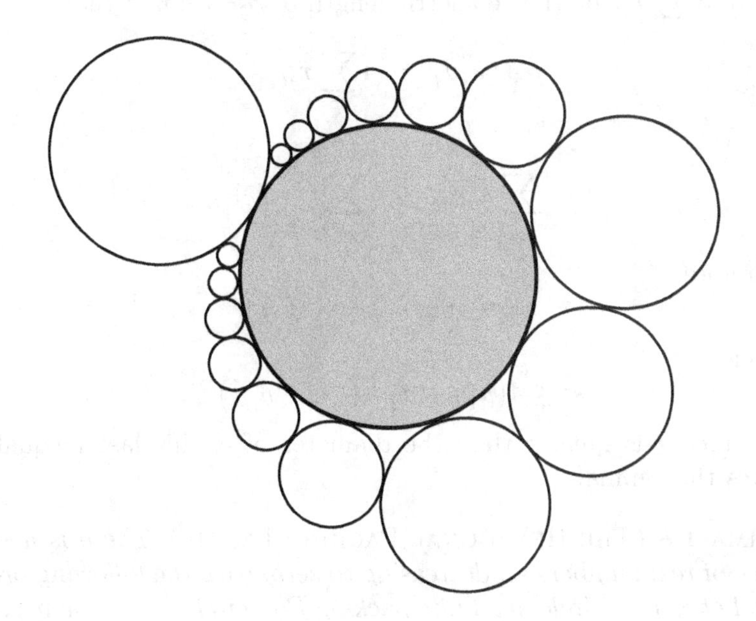

Figure 1.4: The ring lemma.

Proof of the Lemma: Fix k. There is a uniform lower bound for the radius of the largest outer circle c_1 (this occurs when all the k outer circles have equal radius). A circle c_2 adjacent to c_1 also has a uniform lower bound for its radius because, if c_2 were extremely small, then a chain of $k-1$ circles starting from c_2 could not escape from the crevice between c_1 and the unit circle. Repeat this reasoning for the circle c_3 adjacent to c_2, and so forth. □

LEMMA 1.7 (THE LENGTH-AREA LEMMA) *Let c be a circle in a circle packing of the unit disc. Let S_1, S_2, ..., S_k be k disjoint chains which separate c from the origin and from a point of the boundary of the disc. Denote the combinatorial lengths of these chains by n_1, n_2, ... n_k. Then*

$$\text{radius}(c) \leq \left(n_1^{-1} + n_2^{-1} + \cdots n_k^{-1}\right)^{-1}.$$

Proof: Suppose that the chain S_j consists of circles of radii $r_{j\ell}$. Then, by the Schwarz inequality,

$$\left(\sum_\ell r_{j\ell}\right)^2 \leq n_j \sum_\ell r_{j\ell}^2.$$

Let $s_j = \sum_\ell r_{j\ell}$ be the geometric length of S_j. We find that

$$s_j^2 n_j^{-1} \leq 4 \sum_\ell r_{j\ell}^2$$

and

$$\sum_j s_j^2 n_j^{-1} \leq 4 \sum_{j,\ell} r_{j\ell}^2 \leq 4.$$

Therefore

$$s \equiv \min\{s_1, s_2, \ldots, s_k\}$$

satisfies

$$s^2 \leq 4\left(n_1^{-1} + n_2^{-1} + \cdots + n_k^{-1}\right)^{-1}.$$

Since s is greater than the diameter of c, this last inequality proves the lemma. □

LEMMA 1.8 (THE HEXAGONAL PACKING LEMMA) *There is a sequence of real numbers s_j, decreasing to zero, with the following property. Let c_0 be a circle in a finite packing P of circles in the plane, and suppose that the packing P around c_0 is combinatorially equivalent* *to n generations of the regular hexagonal circle packing about one of its circles. Then the ratio of radii of any two circles in the flower around c differs from unity by less than s_n.*

COROLLARY 1.9 *A circle packing in the plane with the hexagonal pattern is the regular hexagonal packing.*

Proof of the lemma: Suppose that, for each $n = 1, 2, \ldots$, we have a circle packing P_n which is combinatorially equivalent to n generations of the regular hexagonal circle packing centered around the unit circle c_0. We apply the Ring Lemma to the P_ns. This shows that the radii of circles of generation k in $P_k, P_{k+1}, \ldots$ are uniformly bounded away from zero and infinity. Therefore we can select a subsequence of the P_js so that the generation one circles converge geometrically. A further subsequence can be selected so that the generation two circles converge geometrically, and so forth.

In this fashion, a limit infinite packing of a planar domain is obtained. This packing has the combinatorics of the regular hexagonal packing, and the domain is an increasing sequence of discs and so is connected and simply connected.

If the lemma were false, then we could select the subsequences so that, in the limit packing, one of the six circles around c_0 would have radius different from 1. This contradicts the uniqueness of packings in the plane with the pattern of the hexagonal packing. □

REMARK 1.10 It appears that we are using the corollary to prove the lemma, but an independent proof of the corollary can be given. See Appendix 1 in [ROS].

1.2.4 Convergence of Circle Packings to the Riemann Mapping

Let U be a simply connected, bounded domain in the plane with two distinguished points $z_0, z_1 \in U$. For a small $\epsilon > 0$, consider the regular hexagonal packing H_ϵ of the plane by circles of radius ϵ. Let c_0 be a circle whose flower contains z_0. Form all chains $C_0, C_1, \ldots, C_k$ of circles from H_ϵ, starting with c_0, such that the flowers of the circles in the chain are contained in U. The circles which appear in such chains are called *inner circles*. The set of inner circles is denoted by I_ϵ.

The circles in H_ϵ which are not inner circles but which are tangent to inner circles will be called *border circles*. The set B_ϵ of border circles can be cyclically ordered to form a cycle called the *border*.

The border has the property that the linear polygon obtained by joining the centers in order is a Jordan curve surrounding the inner circles. The inner circles I_ϵ and border circles B_ϵ form a packing C_ϵ whose nerve is the 1-skeleton of a triangulation T_ϵ.

Complete T_ϵ to a topological triangulation T_ϵ^* of the sphere by adding a vertex at ∞ plus disjoint (except for the point at ∞) Jordan arcs from ∞ to the centers of the border circles.

By Andreev's theorem, there is a circle packing of the sphere whose associated triangulation is isomorphic to T_ϵ^* by an isomorphism that preserves the orientation of the sphere. This circle packing is unique up to Möbius transformations. We partially normalize this packing so that the exterior of the unit disc is the disc whose center correspongs to the vertex ∞ of T_ϵ^*. We then have a correspondence $c \to c'$ of circles c in $C_\epsilon \equiv I_\epsilon \cup B_\epsilon$ with circles c' in a circle packing C_ϵ' of the unit disc. We further normalize things by a Möbius transformation fixing the unit disc so that c_0' is centered at the origin and c_1', where c_1 is a circle whose flower contains z_1, is centered on the positive real axis.

The correspondence $c \to c'$ of C_ϵ to C_ϵ' defines an approximate mapping of R into the unit disc D in the following fashion. Let z be any point in U. For $\epsilon > 0$ sufficiently small, z will be in the flower of some c in I_ϵ. As $\epsilon \to 0$, such a flower will be surrounded by more and more generations of cycles in I_ϵ. Therefore, by the Length-Area lemma and the divergence of the harmonic series, the radius of c' shrinks to 0. Thus c' determines an approximate position for the image of z in D.

1.2.5 The Main Result

And now we have our main result:

Prelude: This theorem is the heart of the matter. Now we see where the approximate conformal mapping comes from, and why it converges to the conformal mapping that we seek.

THEOREM 1.11 *The isomorphism $C_\epsilon \to C_\epsilon'$ of circle packings determines an approximate mapping which, as $\epsilon \to 0$, converges to a conformal homeomorphism of U with the unit disc D.*

Proof: As described in our preceding discussion, we have a circle packing C_ϵ in U and an isomorphic circle packing C_ϵ' of the unit disc D. The associated isomorphic triangulations are denoted by T_ϵ and T'_ϵ. Let R_ϵ and R'_ϵ be the carriers of T_ϵ and T'_ϵ. Let $f_\epsilon : R_\epsilon \to R'_\epsilon$ be the simplicial mapping determined by the correspondence of the vertices of T_ϵ and T'_ϵ. We may assume that f_ϵ is orientation preserving.

The construction makes it clear that R_ϵ converges to U in the sense that U is the union of the R_ϵ, and any compact subset of U is contained in all R_ϵ with sufficiently small positive ϵ (this is a special case of Carathéodory domain convergence). It is also true that R'_ϵ converges to the disc D in the same sense; this follows from the Length-Area Lemma which shows that the radii of the border circles of C'_ϵ tend uniformly to 0 as $\epsilon \to 0$ (each border circles is separated from the origin of D by many disjoint chains of combinatorial length $\leq 6, 12, 18, \ldots$).

The Ring Lemma shows that the angles of the triangles in T'_ϵ are bounded away from 0 independently of ϵ. The Ring lemma, in fact, shows this directly for the inner triangles of T'_ϵ. The proof of the Ring lemma can also be applied to border circles to show that the ratio of the radius of a border circle to the radius of any adjacent circle in C'_ϵ is bounded above. It is also bounded away from zero. Hence the maps $f_\epsilon : R_\epsilon \to R'_\epsilon$ are unformly K-quasiconformal since they map equilateral triangles to triangles of uniformly bounded distortion.

Since the f_ϵ are K-quasiconformal, it follows that they are equicontinuous on compact subsets of U (the same is true for the family f_ϵ^{-1} on compact subsets of D). This standard fact can be seen, for example, by observing that, if z, z' vary in a compact subset of U and $|z - z'|$ becomes arbitrarily small, then $|f_\epsilon(z) - f_\epsilon(z')|$ cannot remain bounded away from 0 for for a sequence of ϵ tending to 0. Indeed, one can surround z, z' with annuli A of arbitrarily large modulus; hence $f_\epsilon(A)$ has arbirarily large modulus and surrounds $f_\epsilon(z)$, $f_\epsilon(z')$. This is impossible if $f_\epsilon(z)$, $f_\epsilon(z')$ lie in a bounded domain and their distance apart is bounded away from 0.

From the equicontinuity on compacta, it follows that the f_ϵ form a normal family. From the Carathéodory domain convergence of R_ϵ to U and of R'_ϵ to D it follows that every limit function f is a K-quasiconformal mapping of U onto D with $f(z_0) = 0$ and $f(z_1) > 0$. That $D \supset f(U)$ follows from the fact that $f(z_0) = 0$.

To see that $D = f(U)$ we pick $w_0 \in D$. Let U be a subdomain of D with $0 \in U$, $w_0 \in U$, and $D_\epsilon \subset U$ for all sufficiently small $\epsilon > 0$. Consider now the restrictions of f_ϵ^{-1} to U; denote them by g_ϵ. Choose $\epsilon(n) \to 0+$ such that $f_{\epsilon(n)} \to f$ and $g_{\epsilon(n)} \to g$ uniformly on compacta. Now $U \supset g(U)$ because $g(0) = z_0$. It follows from $f_{\epsilon(n)}(g_{\epsilon(n)}(w_0)) = w_0$, using the uniform convergence of $f_{\epsilon(n)}$ near $g(w_0)$, that $f(g(w_0))=w_0$. Hence $D = f(U)$. Also f is one-to-one since the roles of U and D can be reversed.

The Hexagonal Packing lemma tells us that the simplicial mapping f_ϵ restructed to a fixed compact subset of U maps equilateral triangles to triangles of G'_ϵ that become arbitrarily close to equilateral as $\epsilon \to 0$. Therefore any limit function f of the f_ϵ will be 1-conformal and hence conformal. Since $f(z_0) = 0$ and $f(z_1) > 0$, we see that all limit functions are equal to the unique Riemann mapping with this normalization.

1.3 The Riemann Mapping Theorem by Way of the Green's Function

Capsule: Of course the Green's function is an important artifact of harmonic analysis and complex function theory. The Green's function is used to construct the Poisson kernel, and is fundamental to solving the Laplace equation.

Interestingly, the Bergman kernel of a domain can be expressed in terms of the Green's function (see our treatment in Chapter 9). It follows, therefore, that the Green's function has important invariance properties. It thus comes as no surprise that we can construct a proof of the Riemann Mapping theorem using the Green's function. That is the purpose of this section.

Now fix a bounded domain $U \subseteq \mathbb{C}$. We shall assume that U has twice continuously differentiable boundary—just so that we can make good use of some results about boundary value problems for the Laplacian and associated ideas (see [KRA2, Chapter 1] for further details). Riemann himself also made some regularity assumptions about the boundary. We also assume, at least for the moment, that U is topologically trivial.

The Green's function is a modification, adapted to a given domain, of the fundamental solution for the Laplacian. It is useful in constructing the Poisson kernel, and in other partial differential equations applications.

Fix a point $b \in U$. Let F_b be the solution of the Dirichlet problem with boundary data $z \mapsto (-1/[2\pi]) \log|z-b|$. Define

$$G(b,z) = -\frac{1}{2\pi}\log|z-b| - F_b(z).$$

This is the Green's function for U with pole at b. See [KRA1, Ch. 1] for a detailed consideration of the Green's function.

Now let $\widetilde{G}(b,z)$ be the harmonic conjugate of G on U. That is to say, $z \mapsto G(b,z) + i\widetilde{G}(b,z)$ is holomorphic in z. Define

$$F(z) = e^{-G(b,z)-i\widetilde{G}(b,z)}.$$

Then F is holomorphic and vanishes at b (because of the logarithmic singularity of G at b). Observe that, if $z \in \partial U$, then

$$|F(z)| = \left|e^{-G(b,z)-i\widetilde{G}(b,z)}\right| = \left|e^{-G(b,z)}\right| = e^0 = 1. \tag{1.12}$$

Since the Green's function is positive, $-G$ is negative and hence F maps U *into* the disc.

> **Claim:** The function F is a one-to-one, onto mapping of U to the unit disc D.

Thus F is starting to look like the Riemann mapping!

To verify the claim, first note that $|F(z)| \to 1$ as $z \to \partial\Omega$. The function F has only one zero, a simple zero, at b. If α is any point of the disc, then we may apply the argument principle to conclude that $F(z) - \alpha$ has precisely one zero in Ω. Thus F is both one-to-one and onto.

Just for fun, we now also present an ad hoc argument to see these assertions in a slightly different fashion. By a translation of coordinates, we may take b to be the origin. Now the Riemann Mapping theorem provides a conformal mapping $\varphi : U \to D$ such that $\varphi(0) = 0$. We know that the Green's function for D at 0 is $(-1/[2\pi]) \log |z|$. It follows immediately that $(-1/[2\pi]) \log |\varphi(z)|$ is the Green's function for U at 0. So

$$G(z,0) = -\frac{1}{2\pi} \log |\varphi(z)|.$$

Exponentiating both sides and adding in a conjugate function of $G(z,0)$, we find that

$$\varphi(z) = e^{-G(z,0)-i\widetilde{G(z,0)}}.$$

This equation is quite similar to (1.12). In fact, they differ only by a unimodular multiplicative constant. We conclude that the function F defined above is one-to-one and onto.

This last ad hoc argument is not entirely satisfactory since it *uses* the Riemann Mapping theorem (and we are trying to derive the Riemann Mapping theorem), but it gives us a different perspective.

1.4 The Ahlfors Map

Capsule: Lars Ahlfors had the elegant idea of examing the proof of the Riemann Mapping theorem and seeing what it said about a multiply connected domain. The result of his studies is the Ahlfors map. This is a mapping from an m-connected domain U to the disc that is an m-fold covering map (in the sense of topology).

For many purposes, the Ahlfors map is as good as the Riemann mapping. One can prove theorems on a multiply connected domain by first proving them on the disc and then transferring them to the domain using the Ahlfors map.

There are many approaches to uniformization of multiply connected domains in the plane. In the present section, we will present Ahlfors's ideas that led to the Ahlfors map. For a detailed reference on this matter, see [FIS]. Also [KRA7] has some material on the Ahlfors map.

We fix a bounded domain $U \subseteq \mathbb{C}$ that is finitely connected, that is, its complement has finitely many components. Assume, as usual, that no component of the complement is a singleton. We begin our analysis just as with the proof of the Riemann Mapping theorem (RMT). That is to say, fix a point $P \in U$. Set

$$\gamma = \sup\{|f'(P)| : f \in H^\infty(U), \|f\|_\infty \leq 1\}.$$

Any function $f \in H^\infty(U)$ such that $\|f\| = 1$ and $f'(P) = \gamma$ we shall call *extremal*. Just as in the proof of the RMT, we can prove that an extremal exists with a simple normal families argument. The details are omitted.

PROPOSITION 1.13 *There is just one extremal function* f, *and* $f(P) = 0$.

REMARK 1.14 Observe that, if we defined an extremal by the (perhaps more common) condition $|f'(P)| = \gamma$, then the extremal would no longer be unique (just postcompose with a rotation). It is the positivity of the derivative at P that gives us uniqueness.

Proof of the Proposition: Let f be an extremal function. Define, for $z \in U$,

$$g(z) = \frac{f(z) - f(P)}{1 - \overline{f(P)}f(z)}.$$

Then

$$g'(z) = \frac{f'(z)(1-|f(P)|^2)}{[1-\overline{f(P)}f(z)]^2}$$

and

$$g'(P) = \frac{\gamma}{1-|f(P)|^2}.$$

Since $\|g\|_\infty \leq 1$, we see that the function g is a candidate for the extremal. It must be that $f(P) = 0$ otherwise $g'(P) > \gamma$.

For the uniqueness, suppose that both f_1 and f_2 are extremals. Then certainly $[f_1 + f_2]/2$ is also an extremal. If we can show that any extremal must be an extreme point of the unit ball of $H^\infty(U)$, then it must follow that $f_1 = f_2$ and the proof will be complete.

Thus let $g \in H^\infty$ and suppose that $\|f \pm g\|_\infty \leq 1$. It follows that

$$|f|^2 + 2\text{Re}\, f\overline{g} + |g|^2 \leq 1$$

and

$$|f|^2 - 2\text{Re}\, f\overline{g} + |g|^2 \leq 1.$$

Adding these inequalities together yields

$$|f|^2 + |g|^2 \leq 1.$$

As a result,

$$|g|^2 \leq 1 - |f|^2 = (1+|f|)(1-|f|) \leq 2(1-|f|).$$

Now define

$$h = \frac{g^2}{2}.$$

We see that $h \in H^\infty$ and

$$|f| + |h| \leq |f| + (1-|f|) = 1.$$

We shall prove that this last estimate implies that h vanishes identically. Thus g vanishes identically, and so f is an extreme point of the unit ball of H^∞.

First we claim that $h(P) = 0$. If not, then we examine the function $f + \lambda f h$ for a constant λ of modulus 1. Notice that

$$|f + \lambda f h| \leq |f| + |f||h| \leq |f| + |h| \leq 1.$$

But then

$$[f + \lambda f h]'(P) = f'(P) + \lambda f'(P)h(P) + \lambda f(P)h'(P) = \gamma + \lambda\gamma h(P).$$

If $h(P) \neq 0$, then an appropriate choice of λ makes this last quantity positive and greater than γ, contradicting the maximal value of γ. That establishes the claim.

If h is not identically 0, then h has some finite order of vanishing r at P, $r > 1$. Let $\epsilon > 0$ and set

$$m(z) = f(z) + \epsilon h(z)(z - P)^{-r+1}.$$

Then $|m'(P)| = |\gamma + \epsilon h^{(r)}(P)/r!|$ (where the exponent (r) denotes the rth derivative) and this quantity can be made to exceed γ if we simply choose the argument of ϵ appropriately. Further observe that, if ϵ is small and $|z - P| > \epsilon^{1/[r-1]}$, then $|\epsilon(z - P)^{-r+1}|$ is less than 1. But then

$$|m(z)| \leq |f(z)| + |h(z)| \leq 1$$

for these values of z. By the maximum principle, we see that $|m| \leq 1$, and we have once again contradicted the definition of extremal. In conclusion, $h \equiv 0$, and the proof is complete. □

Definition 1.15 The unique extremal function provided by the last proposition will be called the *Ahlfors map* or *Ahlfors function* of the domain U with respect to the point P.

The Ahlfors function will play the role, for a multiply connected domain, of the Riemann Mapping function for a simply connected domain. In honor of Lars Ahlfors, we shall denote this function by $\alpha(z)$. Of course, just for topological reasons, the Ahlfors function will *never* be one-to-one (unless the domain in question is simply connected). It will instead be a covering map.

Definition 1.16 A subset $S \subseteq U$ is called *dominating* if

$$\sup_{s \in S} |f(s)| = \sup_{z \in U} |f(z)|$$

for all functions $f \in H^\infty(U)$.

LEMMA 1.17 *The domain U has a countable dominating set that has no limit point in U.*

Proof: Let K_j be a sequence of compact subsets of Ω such that

$$K_j \subseteq \overset{\circ}{K}_{j+1}$$

for each j and $\cup_j K_j = \Omega$. Define

$$E_j = K_j \setminus \overset{\circ}{K}_{j-1}, \ j = 2, 3, \ldots.$$

Certainly each E_j is compact. Also, for each $f \in H^\infty(\Omega)$,

$$\max_{E_j} |f(z)| = \max_{K_j} |f(z)| \to \|f\|_\infty$$

as $j \to \infty$.

On each E_j, the unit ball $\mathcal{B}$ in $H^\infty(U)$ forms an equicontinuous family. To see this, notice that there is a number $r > 0$ such that $D(e,r) \subseteq \Omega$ for each $e \in E_j$. Apply the Cauchy estimates to any $f \in \mathcal{B}$ on $D(e,r)$. Now let us select $\delta_j > 0$ such that

$$|f(z) - f(w)| < \left(\frac{1}{2}\right)^j \quad \text{if } |w - z| < \delta_j, \ w, z \in E_j, \ f \in \mathcal{B}.$$

Let T_j be a finite set in E_j such that each point of E_j is distance not greater than $\delta_j/2$ from some element of T_j. [We call T_j a $\delta_j/2$-*net* in E_j.] It follows that

$$\max_{E_j} |f(z)| \leq \max_{T_j} |f(z)| + \left(\frac{1}{2}\right)^j .$$

Now set $S = \cup_j T_j$. This S is the dominating set that we seek. □

REMARK 1.18 It is intuitively clear that dominating sets should exist. As an exercise, write down an explicit dominating set for the unit disc.

PROPOSITION 1.19 *Let α be the Ahlfors map for U with respect to the point $P \in U$. Then, for each $h \in H^\infty(U)$, we have*

$$\|\alpha h\|_\infty = \|h\|_\infty .$$

Proof: Set

$$U' = \{z \in U : \exists h \in H^\infty(U) \text{ such that } |h(z)| > 1 \text{ and } \|\alpha h\| \leq 1\}.$$

Seeking a contradiction, we suppose that $P \in U'$. Set $\alpha_1 = h \cdot \alpha$. Thus $\|\alpha_1\| \leq 1$. But

$$|\alpha_1'(P)| = |\alpha'(P)| \cdot |h(P)| > |\alpha'(P)| > 1,$$

and that is impossible. So $P \notin U'$.

We shall show that U' is open and closed in U. Since the last paragraph shows that U' is not all of U, the only possible conclusion is that $U' = \emptyset$. This will prove the proposition.

It is clear from its very definition that U' is open. We concentrate on showing that the set is closed; in fact we shall show that $U \setminus U'$ is (relatively) open. Fix a point $q \in U \setminus U'$. Then $\|\alpha h\| \leq 1$ implies that $|h(q)| \leq 1$ for all $h \in H^\infty$. Let $\{s_j\}$ be a countable dominating sequence in U with no limit point in U (as provided by Lemma 1.17). We may of course assume that no s_j is equal to q. Let $\mathbf{M}$ be the maximal ideal space of ℓ^∞. Any function w that is bounded on a neighborhood of $\{s_j\}$ gives rise to a continuous function $\widehat{w}$ on $\mathbf{M}$. More precisely, define

$$U_j = w(s_j)$$

for $j = 1, 2, \ldots$. Then $\mathcal{W} = \{U_j\} \in \ell^\infty$ and we have the induced function

$$\widehat{w} : \mathbf{M} \to \mathbb{C}$$

given by

$$\widehat{w}(\mathbf{m}) = \mathcal{W}(\mathbf{m}) \qquad \text{for } \mathbf{m} \in \mathbf{M}.$$

In particular, if $w \in H^\infty(U)$, then w induces a continuous function on $\mathbf{M}$.

Now our hypothesis that $q \notin U'$ implies that the linear functional from αH^∞ into $\mathbb{C}$ given by

$$\alpha h \mapsto h(q)$$

has norm 1. The Hahn-Banach theorem now yields that there is a measure μ on $\mathbf{M}$, having norm (i.e., total mass) 1, such that

$$\int \widehat{\alpha}\,\widehat{h}\, d\mu = h(q) \quad \text{for all } h \in H^\infty(U).$$

Therefore

$$1 = \int \widehat{\alpha}\,\widehat{1}\, d\mu \leq \int |\widehat{\alpha}||d\mu| \leq \|\mu\| = 1.$$

We conclude that the measure $d\tau = \widehat{\alpha}\, d\mu$ is nonnegative and has mass 1; moreover, τ is supported on the set where $|\widehat{\alpha}| = 1$.

For $\zeta \in U$ a point near q, let

$$s_\zeta(z) = \frac{z-q}{z-\zeta}, \quad z \in U.$$

Set

$$w(\zeta) = \int_U \widehat{s}_\zeta(z)\, d\tau(z).$$

Clearly the function w is continuous at points near q. Also, since $w(q) = 1$, we may conclude that $w(\zeta) \neq 0$ if ζ is near to q. Let ζ be such a point.

If $h \in H^\infty(U)$, then we define

$$g(z) = \frac{h(z) - h(\zeta)}{z - \zeta} \cdot (z - q).$$

Then

$$0 = \int \widehat{g}\, d\tau$$

because $g(q) = 0$. Therefore

$$h(\zeta)u(\zeta) = \int \widehat{h}\widehat{s}_\zeta\, d\tau,$$

and this implies that

$$|h(\zeta)| \leq C \cdot \sup\{|\widehat{h}(\mathbf{m})| : \mathbf{m} \in \mathbf{A}\},$$

where $\mathbf{A}$ is the support of τ and C is a constant that does not depend on h.

Now we replace h by h^j, take the jth root of both sides, and let $j \to \infty$. We conclude that

$$|h(\zeta)| \leq \|\widehat{h}\|_{\mathbf{A}} = \sup\{|\widehat{h}(\mathbf{m})| : \mathbf{m} \in \mathbf{A}\}.$$

The upshot is that, if $\|hF\| \leq 1$, then $|\widehat{h}| \leq 1$ on $\mathbf{A}$ and hence $|h(\zeta)| \leq 1$. But then $\zeta \notin U'$. So q has a neighborhood that does not lie in U'. Hence U' is closed. That completes the proof. □

Definition 1.20 Let U be a domain as usual. A point $x \in \partial U$ is said to be *essential* if there are a bounded, holomorphic function f on U and an $\epsilon > 0$ such that f does not extend analytically to be holomorphic on the disc $D(x, \epsilon)$.

If a point $x \in \partial U$ is not essential, then it is said to be *removable.* If all boundary points of a given domain U are essential, then the domain is said to be *maximal.*[1]

In the rest of our exposition about the Ahlfors mapping, we shall assume that ∂U consists of finitely many smooth, closed curves. This extra hypothesis will allow us to focus on the main analytic points and not get bogged down in technical details. Note, in particular, that every boundary point of such a domain is essential. For if $x \in \partial U$,

[1]In the subject of several complex variables these ideas relate to the concept of "domain of holomorphy."

then there is a conformal mapping φ of U to the disc (not, in general, onto the disc) such that the boundary curve containing x goes univalently under φ to the circle and x goes to 1. Since $\lambda(z) = e^{-(1+z)/(1-z)}$ is a bounded, holomorphic function on D that does not analytically continue past 1, we see that $\lambda \circ \varphi$ is a bounded, holomorphic function on U that does not analytically continue past x. Likewise, we may pull the peak function $p(z) = (1+z)/2$ back from D to get a peaking function on U at x.

PROPOSITION 1.21 *Let α be the Ahlfors map for the domain U with respect to the point $P \in U$. For each point $x \in \partial U$, we have*

$$\limsup_{U \ni z \to x} |\alpha(z)| = 1. \tag{1.21.1}$$

Proof: As already noted, $x \in \partial U$ is essential. Assume that (1.21.1) fails. In particular, say that

$$\limsup_{U \ni z \to x} |\alpha(z)| = 1 - \delta$$

for some $\delta > 0$. As noted in the discussion immediately preceding the proposition, there exists an $h \in H^\infty(U)$ such that

$$\limsup_{U \ni z \to x} |h(z)| = 1$$

while

$$\limsup_{U \ni z \to y} |h(z)| < 1$$

for some $y \in \partial U$, $y \neq x$.

Now consider αh. At the point x, we have

$$\limsup_{U \ni z \to x} |\alpha(z)h(z)| \leq 1 - \delta.$$

But, at $\partial U \ni y \neq x$, we have

$$\limsup_{U \ni z \to y} |\alpha(z)h(z)| < 1$$

since h has boundary limit less than 1 while α has boundary limit not exceeding 1. In conclusion, $\|\alpha h\|_\infty < 1 = \|h\|_\infty$. This contradicts Proposition 1.19. In conclusion, for any $x \in \partial U$, (1.21.1) holds. □

Our main result about the Ahlfors map is Theorem 1.23 below; it gives a representation of a multiply connected domain onto the unit disc. Before presenting that result, we must consider one more general fact about the Ahlfors map.

In what follows, a set E in the Riemann sphere $\widehat{\mathbb{C}}$ is called a (Painlevé) null set if the only bounded, holomorphic functions on $\widehat{\mathbb{C}} \setminus E$ are constants (see our detailed treatment in Chapter 8 on analytic capacity). As an example, a singleton is a null set.

LEMMA 1.22 *Let α be the Ahlfors map for U with respect to the point $P \in U$. Let $X \subseteq D$ be those points of the unit disc that are not in the range of α. Then $X_r \equiv X \cap \overline{D}(0,r)$ is a Painlevé null set for $r < 1$.*

The upshot of this last result is that the Ahlfors map is onto. The details of this assertion will follow below. The technical statement of the lemma is just for convenience in proving the result.

Proof of the Lemma: If the conclusion fails for some $r < 1$, then let $D' = D \setminus X_r$ and let β be the Ahlfors map for D' with respect to the point 0. Then $\beta'(0) > 1$; otherwise $\beta'(0) = 1$ and therefore uniqueness of the Ahlfors map implies that $\beta(z) \equiv z$. This would contradict Proposition 1.13.

Now we see that

$$(\beta \circ \alpha)'(P) = \beta'(\alpha(P)) \cdot \alpha'(P) = \beta'(0) \cdot \alpha'(P) > \alpha'(P).$$

Certainly this estimate contradicts the extremality of the Ahlfors map α. The result is proved. □

For us, the main result about the Ahlfors map is contained in the next theorem.

Prelude: This result tells us exactly the geometric nature of the Ahlfors map. In particular, it is an $(m+1)$-fold covering of the unit disc D that extends analytically across the boundary.

THEOREM 1.23 *Let the domain U be bounded by $m+1$ disjoint, real analytic Jordan curves $\gamma_0, \gamma_1, \ldots, \gamma_m$ and let α be the Ahlfors function for U relative to a point $P \in U$. Then*

(a) *The mapping α maps U onto D precisely $m+1$ times;*

(b) *The mapping α continues analytically across each boundary curve γ_j and maps each γ_j diffeomorphically onto the circle;*

(c) *The derivative α' does not vanish on any boundary curve γ_j.*

REMARK 1.24 You can now see explicitly that, when the domain has connectivity $(m+1)$, then the Ahlfors map is an $(m+1)$-fold covering of the disc. It has the additional attractive feature that it extends nicely to the boundary.

Proof of the Theorem: Let η denote ∂U. We let $A(U)$ denote the continuous functions on $\overline{U}$ that are holomorphic on U. Define

$$\sigma = \sup\{|f'(P)| : f \in A(U), \|f\| \leq 1\}.$$

By the Hahn-Banach theorem and the Riesz representation theorem, there is a finite Borel measure μ on ∂U of total variation σ such that

$$\int_\eta f(z)\, d\mu(z) = f'(P) \quad \text{for all } f \in A(U).$$

Let $G(z,P)$ be the Green's function on U having pole at P. Let $\widetilde{G}(z,P)$ be the harmonic conjugate of the Green's function, and set $q(z) = G(z) + i\widetilde{G}(z)$. We know from Propositions I and II in the appendix at the end of the section that

$$iq'(z)\, dz = dU_P(z) \quad \text{for } z \in \eta.$$

Here U_P is the harmonic measure (what we usually call $U(P,U,\partial U)$) for ∂U with respect to P (see Chapter 6 of [KRA7]). In addition, we know that

$$q'(z) = -\frac{1}{z-P} + w(z),$$

where w is a function holomorphic in a neighborhood of $\overline{U}$. As a result,

$$\frac{i}{2\pi}\oint_\eta f(z) q'(z)\, dz = f(P) \quad \text{for all } f \in A(U).$$

We conclude that

$$0 = \oint_\eta f(z)\,(z-P)\left[\frac{i}{2\pi}\frac{q'(z)}{z-P}\, dz - d\mu(z)\right] \quad \text{for all } f \in A(U).$$

Now the F. and M. Riesz theorem essentially implies that the expression in brackets, thought of as a measure, is absolutely continuous with respect to dz (see [FIS, page 85] for a more detailed treatment). Thus, by subtraction, the same is true for $d\mu$. Because $q' \neq 0$ on η, we may write

$$d\mu(z) = r(z)\frac{i}{2\pi}q'(z)\,dz \quad \text{for } z \in \partial U.$$

Here $f \in L^1(\eta, ds)$.

We conclude that

$$0 = \oint_\eta f(z)\,(z-P)\left[\frac{1}{z-P} - r(z)\right]\frac{i}{2\pi}q'(z)\,dz \quad \text{for all } f \in A(U).$$

Now a version of the F. and M. Riesz theorem (see [FIS, page 91]) tells us that

$$[1-(z-P)r(z)]q'(z) = h(z)$$

for $z \in \overline{U}$ and some $h \in H^1(U)$. An equivalent formulation of this last identity is

$$r(z) = \frac{1}{z-P} + \sum_{j=1}^{m} \frac{c_j}{z-t_j} + g(z)$$

for $z \in \overline{U}$. Here $g \in H^1(U)$, the $c_1, \ldots, c_m$ are complex scalars, and $t_1, \ldots, t_m$ are the critical points of q, that is, the zeros of q'. Note that $t_1, \ldots, t_m$ come from the Green's function, and this is where the connectivity of U plays a role.

If α is the Ahlfors mapping for U relative to P, then

$$\begin{aligned}\sigma \equiv \alpha'(P) &= \oint_\eta \alpha(z)r(z)\frac{i}{2\pi}q'(z)\,dz \\ &\leq \oint_\eta |\alpha(z)||r(z)|\left|\frac{i}{2\pi}q'(z)\right|\,d|z| \leq \|\alpha\|_\infty\,\sigma \\ &= \sigma.\end{aligned}$$

As a result,

$$\alpha(z)r(z) \geq 0 \quad \text{a.e. } |dz| \text{ on } \eta \tag{1.23.1}$$

and

$$|\alpha(z)| = 1 \quad \text{a.e. on the set where } r \neq 0. \tag{1.23.2}$$

Now each point z on η has a neighborhood V such that $V \cap U$ is conformally equivalent to the unit disc D and r is in $H^1(V \cap U)$. Thus, by a result in the appendix to this section, we may conclude that both α and r may be analytically continued over each point of η. (This argument requires the canonical factorization, a concept not covered in this book. See however [FIS, page 75], [HOF], and [KRA2, Chapter 8].) Certainly $|\alpha| = 1$ at every point of η. Thus the meromorphic function $r(z)\alpha(z)$ is real and positive on η, hence has as many poles as zeros in U. Note that the function has exactly $m+1$ poles and therefore α has at most $m+1$ zeros in U.

Next we show that $\arg \alpha$ is locally increasing on η. Fix a point $x \in \partial U$. Let $w = \log|\alpha|$ near x. Then $\widetilde{w} = \arg \alpha$ and the Cauchy-Riemann equations yield that

$$0 \leq \frac{\partial w}{\partial \nu} = \frac{\partial \widetilde{w}}{\partial \tau} \quad \text{on } \eta.$$

This is what we wanted to see. Since $\arg \alpha$ must increase on each γ_j by an integer multiple of 2π, we find that α must have at least $m+1$ zeros in U. In conclusion, α has precisely $m+1$ zeros in U and $\arg \alpha$ increases by 2π on each arc γ_j. Now the argument principle tells us that α is $(m+1)$-to-one.

That completes the proof of the theorem. □

APPENDIX to Section 1.4

In this appendix, we collect some technical results that were used in Section 1.4.

Proposition I: Let U be smoothly bounded. Let U_P be harmonic measure for $\Gamma = \partial U$ with respect to P. As usual, let $G(z, P)$ be the Green's function, $\widetilde{G}(z, P)$ its harmonic conjugate, and $q(z, P) = G(z, P) + i\widetilde{G}(z, P)$. Then

$$dU_P(\zeta) = \frac{i}{2\pi} q'(\zeta)\, d\zeta.$$

Proof: This is a familiar calculation. Or see [FIS, page 23] for the full details. □

Proposition II: Let U be a smoothly bounded domain each of whose boundary arcs is in fact real analytic. Let $\Gamma = \partial U$ consist of $m+1$

disjoint, analytic, simple closed curves (in other words, U has connectivity $m+1$). Fix a point $P \in U$. Let $G(z,P)$ be the Green's function as usual, and let $\widetilde{G}(z,P)$ be the (multiple-valued) harmonic conjugate of g as usual. Set $q(z) = G(z,P) + i\widetilde{G}(z,P)$. Then

(a) q' does not vanish on Γ.

(b) q' has exactly m zeros in U, counting multiplicities.

Proof: Certainly q' has a single pole of order 1 at P; this is by the construction of the Green's function. Furthermore, from Proposition I above, we know that $iq'(z)\,dz$ is a nonnegative measure on Γ. As a result, the total change in the quantity

$$\frac{1}{2\pi}\arg q'(z) = \frac{1}{2\pi i}[iq'(z)],$$

as the point z traverses Γ just once, is precisely $m-1$ (because each boundary curve that bounds a bounded component of the complement contributes $+1$—there are m of these—and the outside curve that bounds the unbounded component of the complement contributes -1). But, by the argument principle, this number must be precisely the number of zeros of q' minus the number of poles. That quantity is then $m-1$. So q' has precisely m zeros, and part **(b)** is proved.

For part **(a)**, note that the Cauchy-Riemann equations imply that $\widetilde{g}$ is increasing locally on Γ; hence q is locally one-to-one in a neighborhood of each point of Γ. So certainly q' cannot vanish. □

1.5 Canonical Representations for Multiply Connected Domains

Capsule: It is natural to wonder whether there is a version of the Riemann Mapping theorem for multiply connected planar domains. Of course the target domain cannot be the disc. What can it be?

In this section we begin to answer that question. There are various slit domains that we use instead of the disc. And the method of proof for our canonical representation is also new.

The end result is an appealing and useful representation for finitely connected planar domains.

Any simply connected domain, except the plane, can be conformally represented as the unit disc. What can be said for domains of higher connectivity?

For a doubly connected domain (i.e., a domain whose complement has two connected components, a bounded component and an unbounded component), there is a remarkable and elegant result.

Prelude: This result is a "toy" version of the full result that we shall prove below. Namely, that any finitely connected domain is conformaly equivalent to a slit domain (or a disc with finitely many smaller discs removed). It is already a lovely and useful statement.

THEOREM 1.24 *Let $U \subseteq \mathbb{C}$ be a doubly connected domain such that no component of the complement is a point. Then U is conformally equivalent to an annulus.*

We will, in fact, prove something more general. The ideas here go back to Koebe and others. In what follows, we let $\widehat{\mathbb{C}}$ denote the Riemann sphere, or $\mathbb{C} \cup \{\infty\}$.

Definition 1.25 Let $U \subseteq \mathbb{C}$ be a domain. We say that U is *finitely connected* if its complement $\widehat{\mathbb{C}} \setminus U$ has finitely many connected components. We say that U is k-connected if $\widehat{\mathbb{C}} \setminus U$ has k components.

Prelude: This is the result that we anticipated in the last prelude. We note once again that the target domain can be taken to be a larger disc with $k-1$ smaller discs removed, or it can be the plane with $k-1$ vertical slits.

THEOREM 1.26 *Let $U \subseteq \mathbb{C}$ be a k-connected domain, $k \geq 2$, and assume that no component of the complement is a point. Then there is a one-to-one, onto, holomorphic map $\varphi : U \to \widetilde{U}$, where $\widetilde{U}$ is an annulus $A = \{\zeta \in \mathbb{C} : c < |\zeta| < C\}$ with $k-2$ concentric arcs (lying on circles $|\zeta| = c_1$, $|\zeta| = c_2, \ldots, |\zeta| = c_{k-2}$) removed. See Figure 1.5.*

Our proof will proceed in several steps. It is a nice combination of topology and complex function theory. First we must review some basic ideas about homology, and about periods of integrals. We collect that material in the next section.

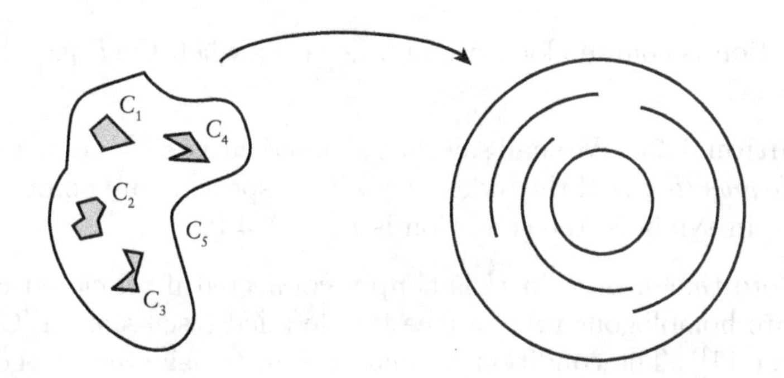

Figure 1.5: An annulus with $k-2$ concentric arcs removed.

1.6 Some Basic Topological Ideas

Capsule: This section develops the topological ideas and the differential forms techniques that will be needed to establish our canonical representation for multiply connected domains. The circle of ideas is reminiscent of de Rham's theorem.

We need the concepts of the index of a point with respect to a curve, of homology, and of periods of integrals. Of course you saw some of these ideas in your complex analysis course. But, for completeness, we need to treat them briefly here.

All curves that we consider in this chapter will be piecewise continuously differentiable. We will not always enunciate this standing hypothesis. A closed curve will be called a *cycle*. In some contexts, it is useful to let a cycle be any formal "sum" of closed curves. (You may find it useful here to compare the presentation with the way that homology is treated in a topology book. See [GRH], [KRA6], and [LEF].)

1.6.1 Cycles and Periods

Definition 1.27 Let γ be a closed curve and a a point not on that curve. Then the *index* of γ with respect to a is

$$n(\gamma, a) = \frac{1}{2\pi i} \oint_\gamma \frac{dz}{z-a}.$$

The index is always an integer (see [AHL2] or [GRK1]). Intuitively, the index measures how many times γ wraps around a, and with what orientation. More precisely, the index is positive when the orientation is counterclockwise and negative when the orientation is clockwise.

Definition 1.28 We shall say that a closed curve γ in a domain U is *homologous to zero* if the index of γ with respect to any point $a \in {}^cU$ is zero. In symbols, the condition is $n(\gamma, a) = 0$.

Note that a domain U is simply connected if all closed curves in U are homologous to zero (see the detailed discussion in [GRK1, Chapter 11]). The condition of a curve γ in U being homologous to zero means that the curve does not "wrap around" any points of the complement.

Now let $U \subseteq \mathbb{C}$ be a multiply connected (but *finitely connected*) domain. This means that the complement of U has at least two but finitely many connected components (see below). Let the components of the complement (in the extended complex plane, or Riemann sphere) be $A_1, A_2, \ldots, A_k$ and suppose that A_k is that component that contains ∞. It is easy to find cycles γ_j, $j = 1, \ldots, k-1$, such that:

- $n(\gamma_j, a) = 1$ for all $a \in A_j$,
- $n(\gamma_j, a) = 0$ for all $a \in {}^cU \setminus A_j$.

See Figure 1.6.

Suppose that γ is *any* cycle in U. Fix an index j, $1 \leq j \leq k$. The value of $n(\gamma, a)$ is, by connectivity, independent of the choice of $a \in A_j$. Call the constant value c_j.

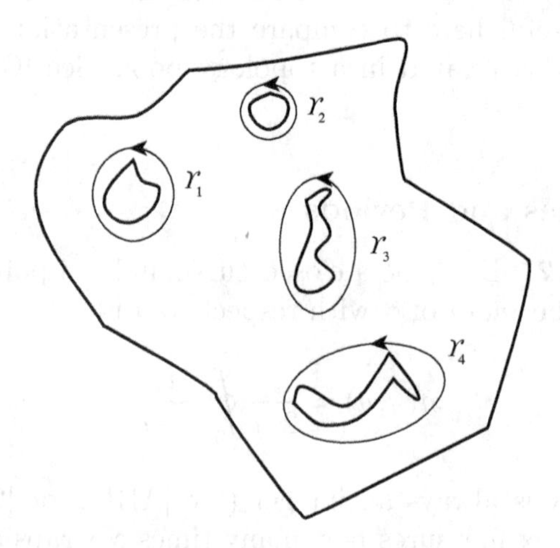

Figure 1.6: A homology basis.

PROPOSITION 1.29 *The cycle*

$$\mu = \gamma - c_1\gamma_1 - c_2\gamma_2 - \cdots - c_{k-1}\gamma_{k-1}$$

has the property that its index with respect to any $a \in {}^cU$ is zero.[2] *(Recall that A_k is the component of the complement that contains ∞, thus it is not considered in this list.) Thus μ is homologous to 0.*

It is convenient to phrase this idea as "the curve γ is homologous to a linear combination of $\gamma_1, \ldots, \gamma_{k-1}$." It is easy to see that the coefficients of this linear combination are uniquely determined.

Definition 1.30 We call $\gamma_1, \ldots, \gamma_{k-1}$ a *homology basis* for U. It is not the only homology basis, but certainly the number of elements in any homology basis will be $k-1$.

We have established that every domain with finite connectivity has a finite homology basis, and vice versa.

Our homology statement has an interpretation in terms of complex line integrals.

PROPOSITION 1.31 *For any function f holomorphic on U and γ a cycle as above we have*

$$\oint_\gamma f(z)\,dz = c_1 \oint_{\gamma_1} f(z)\,dz + c_2 \oint_{\gamma_2} f(z)\,dz + \cdots + c_{k-1} \oint_{\gamma_{k-1}} f(z)\,dz.$$

This is the homology version of Cauchy's theorem.

The numbers

$$P_j \equiv \oint_{\gamma_j} f(z)\,dz$$

are called the periods of the differential $f\,dz$.

As we know from our study of the Cauchy theory (see [AHL2] or [GRK1]), it is possible on a suitable domain (a rectangle $\mathcal{R}$, for instance) to define an antiderivative of a holomorphic function f by fixing a point z_0 in the domain and setting

$$F(z) = \oint_{z_0}^{z} f(\xi)\,d\xi. \tag{1.32}$$

Here the integral is understood to be along *any* piecewise continuously differentiable curve from the fixed base point z_0 to z. Part of

[2] Since the c_j are integers, this notation makes sense. Here $c_j\gamma_j$ means c_j copies of the cycle γ_j.

what one proves—just using Cauchy's theorem—is that the value of the integral is independent of the choice of curve—it only depends on z. Conversely, if f is a given holomorphic function on the rectangle $\mathcal{R}$ and F is a well-defined, single-valued antiderivative for f, then F and f are related (up to an additive constant) by (1.32).

We may put the elementary ideas of the last paragraphs into context in this way:

PROPOSITION 1.33 *On a given domain U, the vanishing of all the periods is a necessary and sufficient condition for f to have a well-defined, single-valued antiderivative. This statement is logically equivalent to the Cauchy integral theorem.*

1.6.2 Harmonic Functions

PROPOSITION 1.34 *If u is a harmonic function on a domain U, that is to say*

$$\triangle u \equiv \left(\frac{\partial^2}{\partial x^2}+\frac{\partial^2}{\partial y^2}\right)u=0,$$

then

$$f(z)=\frac{\partial u}{\partial x}(z)-i\frac{\partial u}{\partial y}(z) \tag{1.34.1}$$

defines a holomorphic function on U.

This claim may be checked directly using the Cauchy-Riemann equations. To wit, let $\alpha=\partial u/\partial x$ and $\beta=-\partial u/\partial y$. Then one sees directly that $\partial\alpha/\partial x=\partial\beta/\partial y$ and $\partial\alpha/\partial y=-\partial\beta/\partial x$.

Now it is useful to write

$$\begin{aligned} f(z)\,dz &= \left(\frac{\partial u}{\partial x}-i\frac{\partial u}{\partial y}\right)(dx+idy) \\ &= \left(\frac{\partial u}{\partial x}\,dx+\frac{\partial u}{\partial y}\,dy\right)+i\left(-\frac{\partial u}{\partial y}\,dx+\frac{\partial u}{\partial x}\,dy\right). \end{aligned}$$

PROPOSITION 1.35 *The real part in this last expression is simply the ordinary, real-variable differential of u:*

$$du=\frac{\partial u}{\partial x}\,dx+\frac{\partial u}{\partial y}\,dy.$$

In the case that u has a conjugate harmonic function v on the domain U, then we must have

$$dv=\frac{\partial v}{\partial x}\,dx+\frac{\partial v}{\partial y}\,dy=-\frac{\partial u}{\partial y}\,dx+\frac{\partial u}{\partial x}\,dy.$$

We know that, in general, even on an annulus, it will not always be the case that u has a well-defined, single-valued harmonic conjugate. So it is best not to discuss or write dv. Instead we introduce the following notation.

Definition 1.36 Let

$$^*du = -\frac{\partial u}{\partial y}\,dx + \frac{\partial u}{\partial x}\,dy.$$

We give $^*du$ the name *conjugate differential of du*. We thus know that

$$f\,dz = du + i\,^*du.$$

This elegant formula gives us a nice way to relate the topology of the domain U to the theory of differential forms on the domain. This is all in the spirit of de Rham's theorem (see [DER]).

Now let γ be any curve that is homologous to zero in U. Then, by Cauchy's theorem,

$$0 = \oint_\gamma f(z)\,dz = \oint_\gamma du + i\oint_\gamma {}^*du.$$

Because du is exact, the first integral on the far right vanishes. So we have

$$\oint_\gamma {}^*du = 0, \tag{1.37}$$

or

$$\oint_\gamma -\frac{\partial u}{\partial y}\,dx + \frac{\partial u}{\partial x}\,dy = 0. \tag{1.38}$$

We wish to give a geometric interpretation of equation (1.38).

It is convenient to write $\gamma(t) = z(t)$ for our closed curve (cycle). If $\alpha = \arg z'(t)$ is the direction of the tangent to γ, then we may write

$$\begin{aligned} dx &= |dz|\cos\alpha, \\ dy &= |dz|\sin\alpha. \end{aligned}$$

Let us consider the normal that points to the right of the tangent. It will have direction $\beta = \alpha - \pi/2$. We see that $\cos\alpha = -\sin\beta$ and $\sin\alpha = \cos\beta$. In summary:

THEOREM 1.39 *The directional derivative*

$$\frac{\partial u}{\partial n} = \cos\beta\cdot\frac{\partial u}{\partial x} + \sin\beta\cdot\frac{\partial u}{\partial y}$$

is the normal derivative of u in the direction of the normal that points to the right of the tangent.

Thus we have

$$\begin{aligned} {}^*du &= -\frac{\partial u}{\partial y}\,dx + \frac{\partial u}{\partial x}\,dy \\ &= -\frac{\partial u}{\partial y}|dz|\cos\alpha + \frac{\partial u}{\partial x}|dz|\sin\alpha \\ &= \left(\sin\beta\frac{\partial u}{\partial y} + \cos\beta\frac{\partial u}{\partial x}\right)|dz| \\ &= \left(\frac{\partial u}{\partial n}\right)|dz|. \end{aligned}$$

So now we may rewrite (1.37) as

$$\int_\gamma \left(\frac{\partial u}{\partial n}\right)|dz| = 0$$

for all cycles γ that are homologous to 0 in U.

Let us repeat what we said in the last subsection using our new language. If U is a simply connected domain, then of course *all* cycles are homologous to zero and hence $\oint_\gamma {}^*du = 0$ for all cycles γ. But then, as we have noted, u will have a well-defined, single-valued conjugate harmonic function on U. As a result, we have the following.

PROPOSITION 1.40 *If U is multiply connected, then the conjugate function has periods*

$$\oint_{\gamma_j} {}^*du = \int_{\gamma_j} \frac{\partial u}{\partial n}|dz|$$

corresponding to the cycles $\gamma_1, \ldots, \gamma_{k-1}$ in a homology basis.

1.7 Uniformization of Multiply Connected Domains

Capsule: At long last, we give the proof of the canonical representation of multiply connected planar domains. The proof is a nice combination of geometry, topology, differential forms analysis, and complex function theory.

Now we shall use the topological/analytical ideas developed thus far to prove Theorem 1.26. It should be noted that there are a number of variants of this result. Instead of an annulus with arcs removed, the target domain can instead be the plane with vertical slits removed. Or it can be the disc with smaller subdiscs removed. In all cases, the classical result applies to finitely connected domains. See [AHL2] or [GOL] for detailed discussion of these matters. The reference [HES] treats uniformization on a domain U consisting of a disc with smaller subdiscs removed in the case that ${}^cU$ has *countably many components.*

Now let U be a finitely connected domain in $\mathbb{C}$, and assume that the domain has connectivity $k \geq 2$. Let the connected components of the complement of U (in the Riemann sphere $\widehat{\mathbb{C}}$) be called $A_1, A_2, \ldots, A_k$. Let us suppose, as usual, that A_k is the unbounded component of the complement (if we think of the complement as taken in $\widehat{\mathbb{C}}$ instead of $\mathbb{C}$, then A_k is the component that contains the point ∞). We shall assume that none of these components of the complement is a single point. Because certainly a single point will be removable for a conformal mapping (by the Riemann removable singularities theorem).

Now we shall make an important geometrical simplification. Notice that $\widetilde{U} \equiv \mathbb{C} \setminus A_k$ is a simply connected domain. So of course $\widetilde{U}$ can be conformally mapped to the disc by a mapping φ. Under the mapping φ, U itself is mapped to a new k-connected domain, and $A_1, \ldots, A_{k-1}$ are mapped to the bounded, connected components of its complement. (We shall continue to denote the connected components of the complement of the image of φ by $A_1, \ldots, A_k$.) Then A_k must now be $\{\zeta : |\zeta| \geq 1\}$. The unit circle $\{\zeta : |\zeta| = 1\}$, equipped with positive orientation, is now the outer boundary contour of the new (mapped) domain, and we call it C_k.

Let us look at ${}^cA_1$ with respect to the extended plane $\widehat{\mathbb{C}}$. This is a simply connected domain, and we map it to the complement in $\widehat{\mathbb{C}}$ of the unit disc, with ∞ mapping to ∞. Under this new mapping, of course C_k will get mapped to a smooth (indeed, real analytic), closed curve, which we still denote by C_k. So now our domain has *two* smooth (real analytic) boundary curves: C_1 and C_k. Note that the inner contour is the unit circle C_1 with negative orientation. The outer contour is just some real analytic curve.

We may repeat this process $k-2$ more times with the curves $C_2, C_3, \ldots, C_{k-1}$ to obtain a domain bounded by an outer contour C_k and $k-1$ inner contours $C_1, \ldots, C_{k-1}$.

PROPOSITION 1.41 *All these new contours are smooth (indeed, real analytic) and the original domain U is conformally equivalent to this new domain.*

In our construction, each boundary curve is momentarily a circle. But, in the end, we can say only that each boundary curve (except possibly the last one) is a real analytic curve. We give the components of ∂U the usual positive orientation. This means that the "outer boundary curve" (the curve that bounds the unbounded component of the complement) is oriented *counter*clockwise; and the other boundary curves (that bound the bounded components of the complement) are oriented *clockwise*. See Figure 1.7.

It is important to note that the index of any point in the plane with respect to the various contours present is obvious. Assume that it is the kth curve that bounds the unbounded component of the new rendition of our domain. Now, the index of C_j, $j < k$, with respect to any point in the interior of C_j is -1. The index with respect to any point exterior to C_j is 0. The index of C_k with respect to any point that is in A_k (the unbounded component of the complement of U) is 0, and the index with respect to all other points not on C_k is 1. We see that $C_1 + C_2 + \cdots + C_k$ bounds U.

We now have the advantage of working on a finitely connected domain whose boundary curves are all real analytic. So certainly

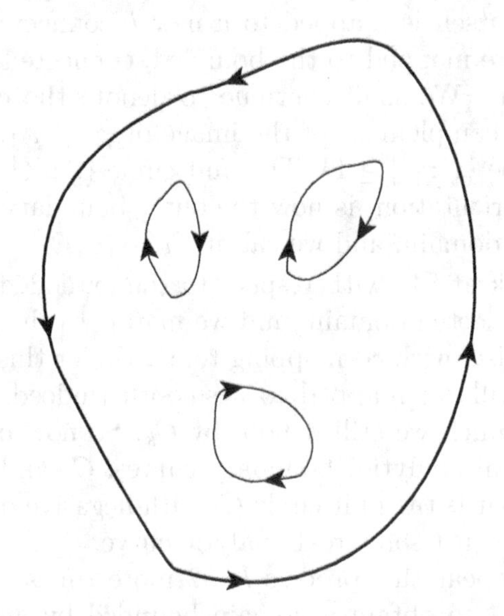

Figure 1.7: Domain bounded by real analytic curves.

every boundary point has a barrier (this term is discussed in detail in [AHL2] or [GRK1]). The Dirichlet problem is therefore solvable (see [AHL2] or [GRK1]). Now, for each $1 \leq j \leq k$, we solve the Dirichlet problem on U with boundary trace equal to 1 on C_j and equal to 0 on the other boundary curves. Call the solution $V_j(z)$. Then $V_j(z)$ is nothing other than harmonic measure (see our Chapter 6 as well as Chapter 6 of [KRA7]) for the curve C_j. Certainly $0 < V_j(z) < 1$ on U and

$$V_1(z) + V_2(z) + \cdots + V_k(z) \equiv 1$$

on U. Notice that V_j can be analytically continued across each boundary curve (by Schwarz reflection—just map each boundary curve to the circle). So we may think of V_j, each j, as harmonic on $\overline{U}$.

The contours $C_1, \ldots, C_k$ form a homology basis for the cycles in U. For each j, the conjugate harmonic differential of V_j has period along C_m given by

$$\alpha_{jm} = \int_{C_m} \frac{\partial V_j}{\partial n}\, ds = \int_{C_m} {}^* dV_j .$$

CLAIM: No linear combination

$$\lambda_1 V_1(z) + \lambda_2 V_2(z) + \cdots + \lambda_{k-1} V_{k-1}(z) \tag{1.42}$$

with constant coefficients λ_j, can have a single-valued harmonic conjugate function *unless all the* λ_j *are zero.*

Proof of the Claim: To verify the claim, suppose instead that (1.42) was the real part of a holomorphic function f on U. By the usual reflection argument, we may suppose that f continues analytically to the closure $\overline{U}$ of U. Then $\operatorname{Re} f$ would take the value λ_j on the contour C_j, $j = 1, \ldots, k-1$, and it would also take the value 0 on C_k. So each of these contours would be mapped under f to a vertical line segment. If τ lies in the complement of all these line segments, then the harmonic function $\arg(f(z) - \tau)$ has a single-valued branch on each contour $C_1, \ldots, C_{k-1}$. The argument principle then tells us that f certainly does not take the value τ in U. But f is a holomorphic function, hence has open image. This is a contradiction. The only possible conclusion is that f is a constant mapping. As a result, the real part of f is identically 0. So the λ_j must all vanish. □

Now we give an interpretation of our result in the language of linear algebra. The homogeneous system of linear equations

$$\begin{aligned}
\lambda_1\alpha_{11}+\lambda_2\alpha_{21}+\cdots+\lambda_{k-1}\alpha_{k-1,1} &= 0,\\
\lambda_1\alpha_{12}+\lambda_2\alpha_{22}+\cdots+\lambda_{k-1}\alpha_{k-1,2} &= 0,\\
&\vdots\\
\lambda_1\alpha_{1,k-1}+\lambda_2\alpha_{2,k-1}+\cdots+\lambda_{k-1}\alpha_{k-1,k-1} &= 0,
\end{aligned}$$

has only the trivial solution $\lambda_1=\lambda_2=\cdots=\lambda_{k-1}=0$, for these are precisely the conditions under which $\lambda_1 U_1+\lambda_2 U_2+\cdots+\lambda_{k-1}U_{k-1}$ has a single-valued, harmonic conjugate. But now this result tells us that any *inhomogeneous system* of linear equations with these same coefficients must, in fact, have a unique solution. In particular, it is certainly possible to solve the system

$$\begin{aligned}
\lambda_1\alpha_{11}+\lambda_2\alpha_{21}+\cdots+\lambda_{k-1}\alpha_{k-1,1} &= 2\pi,\\
\lambda_1\alpha_{12}+\lambda_2\alpha_{22}+\cdots+\lambda_{k-1}\alpha_{k-1,2} &= 0,\\
\lambda_1\alpha_{13}+\lambda_2\alpha_{23}+\cdots+\lambda_{k-1}\alpha_{k-1,3} &= 0,\\
&\vdots\\
\lambda_1\alpha_{1,k-1}+\lambda_2\alpha_{2,k-1}+\cdots+\lambda_{k-1}\alpha_{k-1,k-1} &= 0.
\end{aligned}$$

We append to this system the additional (redundant) equation

$$\lambda_1\alpha_{1k}+\lambda_2\alpha_{2k}+\cdots+\lambda_{k-1}\alpha_{k-1,k}=-2\pi,$$

which is obtained from adding the last $k-1$ equations and using the fact that $\alpha_{j1}+\alpha_{j2}+\cdots+\alpha_{jk}=0$ for each j.

In the language of function theory, the solution of this last system gives us a multiple-valued holomorphic function f with periods $\pm 2\pi$ along C_1 and C_k and all other periods equal to 0. Also, $\operatorname{Re} f$ is constantly equal to λ_j on C_j, $j=1,\ldots,k$ (with $\lambda_k=0$). It follows that the function $F(z)=e^{f(z)}$ is single-valued and holomorphic. We now claim the following result.

Prelude: This is a slightly more precise statement of our main result about multiply connected domains.

THEOREM 1.43 *The function F is a one-to-one, onto, holomorphic mapping from U to the domain $\mathcal{B}$ given by the annulus $\mathcal{A} \equiv \{\zeta \in \mathbb{C} : 1 < |\zeta| < e^{\lambda_1}\}$ with $k-2$ concentric arcs in the circles $\{\zeta \in \mathbb{C} : |\zeta| = e^{\lambda_j}\}$, $j = 2, \ldots, k-1$, removed.*

Proof: We know that F is a well-defined holomorphic function on U and that its image lies in a domain as described in the statement of the theorem. What we must establish are the univalence and surjectivity. Note that the concept of the mapping we are discussing is illustrated in Figure 1.8. Observe again that the contours C_1, C_k correspond respectively to the inner and outer circles of the annulus. The contours $C_2, \ldots, C_{k-1}$ correspond to the arcs that are removed from the annulus. It is useful to think of each arc as a closed cut whose two arc-shaped sides have been forced to coincide.

If $\tau \in \mathbb{C}$, τ not on any of the boundary curves of $\mathcal{B}$, then the number of roots of the equation $F(z) = \tau$ is of course given by

$$\frac{1}{2\pi i}\oint_{C_1}\frac{F'(z)\,dz}{F(z)-\tau}+\frac{1}{2\pi i}\oint_{C_2}\frac{F'(z)\,dz}{F(z)-\tau}+\cdots+\frac{1}{2\pi i}\oint_{C_k}\frac{F'(z)\,dz}{F(z)-\tau}$$
$$\equiv \eta(\tau). \tag{1.43.1}$$

For $\tau = 0$ the terms in (1.43.1) are known to be $1, 0, 0, \ldots, 0, -1$. Hence the sum in (1.43.1) is 0. The integral over C_k (which corresponds to the outer circle of the target annulus) is constantly equal to 1 for $|\tau| < e^{\lambda_1}$; it vanishes for $|\tau| > e^{\lambda_1}$. Likewise the integral over C_1 (which corresponds to the inner circle of the target annulus) is constantly equal to -1 for $|\tau| < 1$ and is 0 for $|\tau| > 1$. Of course the integrals over C_j, $2 \le j \le k-1$, vanish for τ not on the circular arcs.

The aggregate of all this reasoning shows that, if $\tau \in \mathcal{B}$, then the expression in (1.43.1) is the sum of $0 + (0 + \cdots + 0) + 1 = 1$, proving

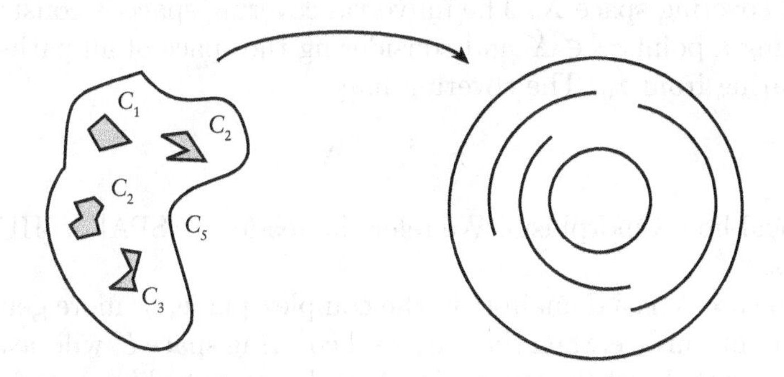

Figure 1.8: Representation of U on an annulus minus arcs.

that the value τ is taken once and only once by the function F. That is what we wished to show. $\square$

We remark that it is possible to calculate the valence of F on the boundary curves. This uses a version of the Plemelj jump formula, and we cannot treat it here. See [AHL2, page 256] for details. Goluzin [GOL] also has a nice treatment of Plemelj.

1.8 The Uniformization Theorem

Capsule: Koebe's uniformization theorem is a very important generalization of the Riemann Mapping theorem. It identifies all the possible covering spaces for Riemann surfaces. The amazing fact is that there are only three: the plane, the sphere, and the disc. Furthermore, the plane only covers the plane, the punctured plane, the torus, and the cylinder. The sphere only covers the sphere. So the vast majority of Riemann surfaces have the disc as the universal covering space.

In other words, most any Riemann surface (except for an explicit, finite list) is the quotient of the disc. This fact gives rise to a detailed and deep structure theory for Riemann surfaces.

We treat the uniformization theorem in this section.

The philosophical godfather of all the theorems discussed in the present chapter is the celebrated uniformization theorem of Poincaré and Koebe. In this section we shall discuss the meaning of the theorem, and then prove a special case of it. The proof hinges on the existence of the Green's function.

If X is any topological space, then it has a simply connected universal covering space $\widehat{X}$. The universal covering space is constructed by fixing a point $x_0 \in X$ and considering the space of all paths in X emanating from x_0. The covering map

$$\pi : \widehat{X} \to X$$

is a local homeomorphism. We refer the reader to [SPA] or [HUS] for details.

In case X is a domain U in the complex plane, or more generally a Riemann surface, then the universal covering space $\widehat{U}$ will be a two-dimensional object (because π is a local homeomorphism), and $\widehat{U}$ can be endowed with a complex structure by local pullback under π of the complex structure from U. So $\widehat{U}$ is a simply connected, analytic object. What is it?

The uniformization theorem answers this question in a dramatic way. Before we present the answer, let us first restate the question—stripped of all the preliminary material that led up to it.

> **QUESTION:** What are all the simply connected Riemann surfaces?

The answer is

> **ANSWER:** The only simply connected Riemann surfaces are
>
> **(i)** the disc D,
>
> **(ii)** the plane $\mathbb{C}$,
>
> **(iii)** the Riemann sphere $\widehat{\mathbb{C}}$.

In fact much more can be said. Let us return to the motivational discussion above. If the original analytic object X is a sphere, then it turns out that the universal covering space $\widehat{X}$ will be a sphere, and that is the *only* circumstance under which a sphere arises as the universal covering space.

If the original analytic object is a plane, a punctured plane, a torus, or a cylinder, then the universal covering space $\widehat{X}$ is a plane, and these are the *only* circumstances in which the plane arises as the universal covering space.

In all other circumstances, the universal covering space is the disc D. In other words,

> The universal covering space for any planar domain, except for $\mathbb{C}$ or $\mathbb{C} \setminus \{0\}$, is the disc D.

This is powerful information, and those who study Riemann surfaces have turned the result into an important tool (see [FAK]). For instance, the result that any automorphism of a planar domain that fixes three points is the identity can be derived as a corollary of the uniformization theorem (see [FKKM]). Many of our other results about automorphisms can be studied with the aid of uniformization.

One way to view the uniformization theorem is that, if U is a planar domain other than the entire plane $\mathbb{C}$ or a punctured plane, then it is a quotient of the disc. Furthermore, if $\mathcal{R}$ is a Riemann surface other than the sphere, a cylinder, or a torus, then it is a quotient of the disc. This is a powerful structure theorem, from which much can be learned.

Our goal in the remainder of this section is to prove the last displayed statement: that the disc is the universal covering space for virtually all planar domains. For convenience and simplicity, we shall restrict attention to bounded planar domains with finitely many smooth boundary curves.

Prelude: This is the most important special case of the uniformization theorem. The plane and the sphere can only be the universal covering space for a very short, finite list of Riemann surfaces. All other Riemann surfaces are covered by the disc.

THEOREM 1.44 *Let $U \subseteq \mathbb{C}$ be a bounded domain whose boundary consists of finitely many simple, closed, C^2 curves. Then the universal covering space for $\widehat{U}$ for U is conformally equivalent to the unit disc D.*

Proof: The important fact for us is that $\widehat{U}$ has a Green's function. This is easily seen by solving a suitable Dirichlet problem. The references [GAM], [KRA2], and [FAK] provide all the details. We shall take it for granted that $\widehat{U}$ has a Green's function. The remainder of the proof presented below studies $\widehat{U}$ and its Green's function.

Fix a point $\xi \in \widehat{U}$ and let $G(z, \xi)$ be the corresponding Green's function. Then, by a construction that we have used before (for instance in Section 1 of this chapter), there is a holomorphic function φ defined near ξ with a simple zero at ξ such that

$$|\varphi(z)| = e^{-G(z,\xi)}.$$

The function φ can be analytically continued along any path in $\widehat{U}$ from ξ to any other point z—just by continuing the harmonic conjugate $\widetilde{g}$ of g and then taking an exponential. Since $\widehat{U}$ is simply connected, the monodromy theorem tells us that the analytic continuation does not depend on the choice of path. By this means, we define a holomorphic function φ on $\widehat{U}$ such that

$$|\varphi(z)| = e^{-G(z,\xi)} \quad \text{for } z \in \widehat{U}.$$

In particular, since G has boundary values 0, we see immediately that $|\varphi(z)| < 1$ for all $z \in \widehat{U}$, and φ has only one zero: the simple zero at ξ.

Fix a point $\tau \in \widehat{U}$ and define

$$\psi(z) = \frac{\varphi(z) - \varphi(\tau)}{1 - \overline{\varphi(\tau)}\varphi(z)} \quad \text{for } z \in \widehat{U}.$$

Then ψ is holomorphic on $\widehat{U}$, $|\psi(z)| < 1$ for $z \in \widehat{U}$, and $\psi(\tau) = 0$. Let u be a subharmonic function on $\widehat{U} \setminus \{\tau\}$ such that

(i) $u = 0$ off some compact subset of U,

(ii) $u(z) + \log|\psi(z)|$ is subharmonic.

Then $u(z) + \log|\psi(z)|$ is subharmonic on $\widehat{U}$ and $u(z) + \log|\psi(z)| < 0$ off a compact subset of $\widehat{U}$. By the maximum principle for subharmonic functions, $u(z) + \log|\psi(z)| < 0$ on all of $\widehat{U}$.

Taking the supremum over all such u, and recalling the construction of the Green's function, we find that

$$G(z,\tau) + \log|\psi(z)| \leq 0 \quad \text{for } z \in \widehat{U}. \tag{1.44.1}$$

Since $\psi(\xi) = -\varphi(\tau)$, we use the symmetry of the Green's function (see [KRA2, Chapter 1]) to determine that

$$G(\xi,\tau) + \log|\psi(\xi)| = G(\xi,\tau) - G(\tau,\xi) = 0.$$

By the strict maximum principle, we see that equality holds in (1.44.1).

Thus $\log|\psi(z)| = -G(z,\tau)$ for all $z \in \widehat{U}$ and so ψ has no zeros on $\widehat{U} \setminus \{\tau\}$. It follows that φ assumes the value $\varphi(\tau)$ only at $z = \tau$. Since τ was quite arbitrary, we deduce that φ is one-to-one. Hence φ maps $\widehat{U}$ conformally onto a domain $\varphi(\widehat{U})$ in the unit disc, and we see that $\varphi(\widehat{U})$ is simply connected. By the Riemann Mapping theorem, $\varphi(\widehat{U})$ is conformally equivalent to the unit disc, and hence so is $\widehat{U}$. □

Chapter 2

Invariant Metrics

Prologue: The first invariant metric in complex variable theory was the Poincaré metric on the unit disc D. This metric has the property that a conformal self-map of the unit disc will preserve distance as measured in the metric.

Once one conceives that such a metric should exist, then it is not difficult to do a calculation to derive the metric explicitly. We leave that task as an exercise for the interested reader. In our treatment, we simply define the metric and then derive its most essential properties.

The interesting development for twentieth-century mathematics is that one can produce an invariant metric for virtually any planar domain, and for many Riemann surfaces. Some of these constructs are inspired by the modern proof of the Riemann Mapping theorem (see the last chapter), and some derive from new ideas connected with Hilbert space.

In turn, the study of invariant metrics has inspired the consideration of complex manifolds, Kähler manifolds, and other geometric constructs. Yau's ideas about Calabi manifolds have hooked up with string theory and given rise to a wealth of new ideas in mathematical physics.

We can consider this chapter to be an introduction to this powerful circle of ideas.

2.1 The Carathéodory Metric

Capsule: Of all the invariant metrics that have been created for planar domains, the Carathéodory metric is one of the most basic. For it is inspired directly by the proof of the Riemann Mapping theorem—particularly by the process of maximizing the derivative at a point.

In addition, it can be shown that the Carathéodory metric is essentially the "smallest" invariant metric (of a certain class) on a domain. Finally, the Carathéodory metric is always majorized by the Bergman metric.

Fix a domain $U \subseteq \mathbb{C}$. Recall that $D \subseteq \mathbb{C}$ is the unit disc.

Definition 2.1 If $P \in U$, define

$$(D,U)_P = \{\text{holomorphic functions } f \text{ from } U \text{ to } D \text{ such that } f(P) = 0\}.$$

The *Carathéodory metric* for U at P is defined to be

$$F_C^U(P) = \sup\{|\varphi'(P)| : \varphi \in (D,U)_P\}.$$

REMARK 2.2 Glancing back at the proof of the Riemann Mapping theorem, especially step **(2)**, we see that the quantity F_C^U measures, for each P, the extreme value posited in the proof of that result. In that proof, it was necessary only to know that the extreme value existed and was finite. Now we shall gain extra information by comparing this value to other quantities.

Clearly $F_C^U(P) \geq 0$. (Moreover, the Cauchy estimates imply that $F_C^U(P) < \infty$.) Is $F_C^U(P) > 0$ for all P? If U is bounded, then the answer is "yes." For then

$$U \subseteq D(0,R) = \{z \in \mathbb{C} : |z| < R\}$$

for some $R > 0$. But then the map

$$\varphi : \zeta \longmapsto \frac{\zeta - P}{2R}$$

satisfies $\varphi \in (D,U)_P$. Thus

$$F_C^U(P) \geq |\varphi'(P)| = \frac{1}{2R} > 0.$$

However, if U is unbounded, then F_C^U may degenerate. If $U = \mathbb{C}$, for instance, then any $f \in (D,U)_P$ is constant, hence $F_C^U \equiv 0$. The same holds for U equaling $\mathbb{C}$ less a discrete set of points (by the

Riemann removable singularities theorem). Which domains have non-degenerate (i.e., non-vanishing) Carathéodory metric (i.e., Carathéodory metric which actually gives a distance)? Analytic capacity is a device for answering this question (see [GAR1]), but we cannot pursue that topic here.

The primary interest of the Carathéodory metric at this moment is that it begins to generalize the Schwarz lemma.

PROPOSITION 2.3 *Let U_1, U_2 be domains in $\mathbb{C}$. Let ρ_j be the Carathéodory metric on U_j. If $h : U_1 \to U_2$ is holomorphic, then h is distance-decreasing from (U_1, ρ_1) to (U_2, ρ_2). In other words,*

$$h^*\rho_2(z) \leq \rho_1(z), \qquad \forall z \in U_1.$$

Here

$$h^*\rho_2(z) \equiv \rho_2(z) \cdot |h'(z)|.$$

COROLLARY 2.4 *If $\gamma : [0,1] \to U_1$ is a piecewise continuously differentiable curve then*

$$\ell_{\rho_2}(h_*\gamma) \leq \ell_{\rho_1}(\gamma).$$

REMARK 2.5 Observe that the Carathéodory metric is not manifestly integrable on curves. But if one thinks about the fact that it is constructed as the supremum of continuous functions, then it follows that the metric is semicontinuous (see [KRA2]). So the metric is indeed integrable on curves (to wit, a semi-continuous function is the monotone limit of continuous functions).

COROLLARY 2.6 *If $P_1, P_2 \in U_1$ then*

$$d_{\rho_2}(h(P_1), h(P_2)) \leq d_{\rho_1}(P_1, P_2).$$

COROLLARY 2.7 *If h is conformal, then h is an isometry of (U_1, ρ_1) to (U_2, ρ_2).*

The corollaries are proved by straightforward reasoning. We concentrate on proving the proposition.

Proof of Proposition 2.3: Fix $P \in U_1$ and set $Q = h(P)$. Notice that if $\varphi \in (D, U_2)_Q$ then $\varphi \circ h \in (D, U_1)_P$. Thus

$$\begin{aligned} F_C^{U_1}(P) &\geq |(\varphi \circ h)'(P)| \\ &= |\varphi'(Q)| \cdot |h'(P)|. \end{aligned}$$

Taking the supremum over all $\varphi \in (D, U_2)_Q$ yields

$$F_C^{U_1}(P) \geq F_C^{U_2}(Q) \cdot |h'(P)|$$

or

$$\rho_1(P) \geq h^*\rho_2(P). \qquad \square$$

Recall that the Poincaré metric on the unit disc, given by

$$g = \frac{1}{(1-|z|^2)^2} dz d\bar{z},$$

has the property that any conformal selfmap of the disc is an isometry of the metric. See [KRA1] for the details.

Next we calculate the Carathéodory metric for the disc.

PROPOSITION 2.8 *The Carathéodory metric on the disc coincides with the Poincaré metric.*

Proof: First we calculate the metric at the origin. If $\varphi \in (D, D)_0$ then, by the Schwarz lemma, $|\varphi'(0)| \leq 1$. Therefore

$$F_C^D(0) \leq 1.$$

But the map

$$\varphi(\zeta) = \zeta$$

satisfies $\varphi \in (D, D)_0$ and $\varphi'(0) = 1$. Thus

$$F_C^D(0) = 1.$$

Because every conformal map of the disc is an isometry of the Carathéodory metric, we may conclude that the Carathéodory metric and the Poincaré metric are equal. $\square$

Given what we've learned so far, we might suspect that any isometry of the Carathéodory metric must be a conformal map. This is indeed true—in fact a much stronger assertion holds—provided the domains in question have nondegenerate (i.e., non-vanishing) Carathéodory metrics. We postpone this topic until after we have introduced the Kobayashi metric.

2.2 The Kobayashi Metric

Capsule: The Kobayashi metric is, in a certain sense, dual to the Carathéodory metric. Its construction is also inspired by the proof of the Riemann Mapping theorem.

The Kobayashi metric is of interest because it majorizes the Carathéodory metric; in fact, the Kobayashi metric is the largest invariant metric of a certain class.

Fix a domain $U \subseteq \mathbb{C}$.

Definition 2.9 If $P \in U$, define

$$(U, D)^P = \{\text{holomorphic functions } f \text{ from } D \text{ to } U \text{ such that } f(0) = P\}.$$

The *Kobayashi* (or *Kobayashi/Royden*) *metric* for U at P is defined to be

$$F_K^U(P) = \inf\left\{\frac{1}{|\varphi'(0)|} : \varphi \in (U, D)^P\right\}.$$

REMARK 2.10 As we remarked in the last section about the Carathéodory metric, the Kobayashi metric quantifies a certain extremal problem and will be useful for comparison purposes. The specific form of the definition of the Kobayashi metric will turn out to facilitate comparisons with the Carathéodory metric; in particular, the two metrics will turn out to interact nicely with the classical Schwarz lemma.

Clearly $F_K^U(P) \geq 0$. In order to learn more about F_K^U, we now compare it with the Carathéodory metric.

PROPOSITION 2.11 *For all $P \in U$,*

$$F_C^U(P) \leq F_K^U(P).$$

Proof: Let $\varphi \in (D, U)_P$ and $\psi \in (U, D)^P$. Then $\varphi \circ \psi : D \to D$ and $\varphi \circ \psi(0) = 0$. It follows that

$$|(\varphi \circ \psi)'(0)| \leq 1$$

or

$$|\varphi'(P)| \leq \frac{1}{|\psi'(0)|}.$$

Taking the supremum over all φ gives

$$F_C^U(P) \leq \frac{1}{|\psi'(0)|}.$$

Now taking the infimum over all ψ yields

$$F_C^U(P) \leq F_K^U(P).$$

□

An immediate consequence of this last result is that, if U is bounded, then F_K^U is nondegenerate, i.e., non-vanishing (for then

F_C^U is). On the other hand, if $U = \mathbb{C}$ then $F_K^U \equiv 0$; for given $P \in \mathbb{C}$, the function

$$\varphi_R(\zeta) = P + R\zeta$$

satisfies $\varphi_R \in (U, D)^P$, any $R > 0$. Thus

$$F_K^U(P) \le \frac{1}{|\varphi_R'(0)|} = \frac{1}{R} \, ;$$

letting $R \to \infty$, we have $F_K^U(P) = 0$.

In analogy with Proposition 2.3 of the last section we now have the following.

PROPOSITION 2.12 *Let U_1, U_2 be domains equipped with the Kobayashi metric (denoted, respectively, by ρ_1 and ρ_2). If $h : U_1 \to U_2$ is holomorphic then h is distance-decreasing from (U_1, ρ_1) to (U_2, ρ_2). That is,*

$$h^*\rho_2(z) \le \rho_1(z), \qquad \forall z \in U_1.$$

COROLLARY 2.13 *If $\gamma : [0, 1] \to U_1$ is a piecewise continuously differentiable curve, then*

$$\ell_{\rho_2}(h_*\gamma) \le \ell_{\rho_1}(\gamma).$$

REMARK 2.14 Just as for the Carathéodory metric, the Kobayashi metric will be integrable on curves because it is semicontinuous. Again see [KRA2].

COROLLARY 2.15 *If $P_1, P_2 \in U_1$ then*

$$d_{\rho_2}(h(P_1), h(P_2)) \le d_{\rho_1}(P_1, P_2).$$

COROLLARY 2.16 *If h is conformal, then h is an isometry of (U_1, ρ_1) to (U_2, ρ_2).*

Proof of Proposition 2.12: Fix $P \in U_1$ and put $Q = h(P)$. Choose $\varphi \in (U_1, D)^P$. Then $h \circ \varphi \in (U_2, D)^Q$. We have

$$\begin{aligned} F_K^{U_2}(Q) &\le \frac{1}{|(h \circ \varphi)'(0)|} \\ &= \frac{1}{|h'(P)||\varphi'(0)|}. \end{aligned}$$

Taking the infimum over all $\varphi \in (U_1, D)^P$ gives

$$F_K^{U_2}(Q) \le \frac{1}{|h'(P)|} \cdot F_K^{U_1}(P)$$

or

$$(h^* F_K^{U_2})(P) \leq F_K^{U_1}(P). \qquad \square$$

Using Proposition 2.12, we may give another example of domains $U \subseteq \mathbb{C}$ with degenerate Kobayashi metric. Let $\mathbb{C}_0 = \mathbb{C} \setminus \{0\}$. Then

$$\begin{aligned} h : \mathbb{C} &\to \mathbb{C}_0 \\ \zeta &\mapsto e^\zeta \end{aligned}$$

is holomorphic. The map h will be distance-decreasing in the Kobayashi metric. Since F_K is identically 0 on $\mathbb{C}$ (because there are arbitrarily large analytic discs in $\mathbb{C}$ centered at any point), it follows that F_K is identically zero on $\mathbb{C}_0$.

As an exercise, prove that we cannot continue this line of reasoning to see that F_K^U is degenerate when $U = \mathbb{C} \setminus \{0, 1\}$. In fact, we shall later see that F_K^U is nondegenerate on $U \equiv \mathbb{C} \setminus \{P_1, \ldots, P_k\}$ provided the P_j's are distinct and $k \geq 2$ (notice that this situation is in contrast to that for the Carathéodory metric as discussed in Section 1). Next we turn our attention to more elementary questions regarding the Kobayashi metric.

PROPOSITION 2.17 *For $U = D \subseteq \mathbb{C}$, the Kobayashi metric equals the Poincaré metric.*

Proof: For $P \in D$ we have

$$F_K^D(P) \geq F_C^D(P) = \rho(P),$$

where ρ is the Poincaré metric on the disc. For an inequality in the opposite direction, first take $P = 0$. Let $\varphi \in (D, D)^0$ be given by

$$\varphi(\zeta) = \zeta.$$

Then

$$F_K^D(0) \leq \frac{1}{|\varphi'(0)|} = 1 = \rho(0).$$

It follows that

$$F_K^D(0) = 1 = \rho(0).$$

We conclude that

$$F_K^D(P) = \rho(P), \qquad \forall P \in D. \qquad \square$$

The first major result of this chapter is as follows (this result is a metric version of the Riemann Mapping theorem):

Prelude: This result begins to suggest why the Carathéodory and Kobayashi metrics may be useful. Any geometric characterization of the unit disc D is a priori of interest.

THEOREM 2.18 *Let $U \subseteq \mathbb{C}$ be a domain. The domain U is conformally equivalent to the disc if and only if there is a point $P \in U$ such that*

$$F_C^U(P) = F_K^U(P) \neq 0.$$

Proof: If U is conformally equivalent to the disc, then let

$$h : U \to D$$

be a conformal map. For any point $P \in U$, we have that

$$F_C^U(P) = (h^* F_C^D)(P)$$

which

$$= (h^* \rho)(P);$$

but this

$$= (h^* F_K^D)(P).$$

We conclude that this last expression equals $F_K^U(P)$, proving the easier half of the theorem.

For the converse, choose $\varphi_j \in (D, U)_P$ such that

$$|\varphi_j'(P)| \to F_C^U(P)$$

and choose $\psi_j \in (U, D)^P$ such that

$$\frac{1}{|\psi_j'(0)|} \to F_K^U(P).$$

Since $\{\varphi_j\}$ is uniformly bounded above by 1, we may extract a subsequence $\{\varphi_{j_k}\}$ converging to a normal limit φ_0. Consider

$$h_{j_k} = \varphi_{j_k} \circ \psi_{j_k}.$$

Passing to another subsequence, which we denote by h_{j_ℓ}, we may suppose that h_{j_ℓ} converges normally to a limit h_0. Notice that $h_0(0) = 0$

so that h_0 is not a unimodular constant; therefore h_0 maps D to D. After renumbering, we call this last sequence

$$h_\ell = \varphi_\ell \circ \psi_\ell.$$

Recall that when a sequence of holomorphic functions converges normally, then so does the sequence of its derivatives. It follows that

$$\begin{aligned} |h_0'(0)| &= \lim_{\ell\to\infty} \left|(\varphi_\ell \circ \psi_\ell)'(0)\right| \\ &= \lim_{\ell\to\infty} |\varphi_\ell'(P)| \cdot |\psi_\ell'(0)| \\ &= F_C^U(P) \cdot \frac{1}{F_K^U(P)} \\ &= 1. \end{aligned}$$

By Schwarz's lemma, $h_0(\zeta) = \mu \cdot \zeta$ for some unimodular constant μ. Thus we have

$$\mu \cdot \zeta = h_0(\zeta) = \lim_{\ell\to\infty} (\varphi_\ell \circ \psi_\ell(\zeta)). \tag{2.18.1}$$

After composing the functions φ_ℓ with a rotation, we may assume that $\mu = 1$.

Now $\mathbb{C} \setminus U$ must contain at least two points (else F_K^U would be identically 0, contradicting our hypothesis). Hence $\{\psi_\ell\}$ forms a normal family. Say that a subsequence ψ_{ℓ_m} converges to some ψ_0. After renumbering, we rewrite (2.18.1) as

$$\begin{aligned} \zeta &= h_0(\zeta) \\ &= \lim_{m\to\infty} (\varphi_m \circ \psi_m(\zeta)) \\ &= \varphi_0 \circ \psi_0(\zeta). \end{aligned} \tag{2.18.2}$$

Since h_0 is surjective, we conclude that φ_0 is surjective.

Because of (2.18.2), the function ψ_0 is a nonconstant holomorphic function; hence its image is open. We claim that the image is also closed (in the relative topology of U). To see this, let $\psi_0(\zeta_j)$ be elements of that image converging to a limit point $q \in U$. Since φ_0 is continuous, it follows that $\varphi_0(\psi_0(\zeta_j))$ converges to a limit point r. But then (2.18.2) tells us that $\zeta_j \to r$. It follows from the continuity of ψ_0 that $\psi_0(r) = q$, so q is in the image of ψ_0. We thus have that the image of ψ_0 is both open and closed, and it is nonempty. Since U is connected, it follows that the image of ψ_0 equals U; hence ψ_0 is surjective. Thus, since h_0 is injective, so must φ_0 be injective. We conclude that φ_0 is the desired conformal map of U to D. □

Notice that the conclusion of the theorem is false if we allow the metrics to degenerate. For example, if $U = \mathbb{C} \setminus 0$, then $F_C^U = F_K^U \equiv 0$; yet U is definitely not conformally equivalent to the disc.

As anticipated in the last section, we now prove that the only isometries of the Carathéodory or Kobayashi metrics which fix a point are conformal maps. In fact, we shall prove something much stronger.

Prelude: Refer to the last prelude. This is a generalization of that result for the disc to a more general class of domains. It is quite common to refer to domains with nondegenerate Kobayashi metric as *Kobayashi hyperbolic*. In a natural sense, Kobayashi hyperbolic domains are an invariant version of bounded domains.

THEOREM 2.19 *Let $U \subseteq \mathbb{C}$ be a domain on which the Kobayashi metric is nondegenerate (i.e., gives a genuine distance function) and fix $P \in U$. Suppose that*

$$f : U \longrightarrow U$$

is a holomorphic function such that $f(P) = P$. Assume that f is an isometry of either the Carathéodory metric or the Kobayashi metric at P; that is, suppose that either

$$f^* F_C^U(P) = F_C^U(P)$$

or

$$f^* F_K^U(P) = F_K^U(P)$$

and that the metric does not vanish at P. Then f is a conformal map of U onto U.

REMARK 2.20 This theorem is a remarkable rigidity statement: the global condition that f map U to U, together with the differential condition at P, forces f to be one-to-one and onto.

Proof of the Theorem: Since U has nondegenerate Kobayashi metric, $\mathbb{C} \setminus U$ must contain at least two points. This implies that any family of holomorphic functions $\{g_\alpha\}$ taking values in U is normal (by a result in the next chapter). We will use this observation repeatedly.

Now the hypothesis that

$$f^* F_C^U(P) = F_C^U(P) \quad \text{or} \quad f^* F_K^U(P) = F_K^U(P)$$

implies that

$$|f'(P)| = 1.$$

Define

$$\begin{aligned} f^1 &= f \\ f^2 &= f \circ f \\ &\vdots \\ f^j &= f^{j-1} \circ f, \qquad j \geq 2. \end{aligned}$$

Then $\{f^j\}$ is a normal family. Since the numbers $(f^j)'(P)$ all have unit modulus (note that $|(f^j)'(P)| = [f'(P)]^j$), there is a subsequence $\{f^{j_\ell}\}$ such that $(f^{j_\ell})'(P) \to 1$. (Exercise: If $f'(P)$ has argument which is a rational multiple of π, then the assertion is clear; if it is an irrational multiple of π, then the set of $(f^j)'(P)$ forms a dense subset of the circle.) Passing to another subsequence, which we still denote by $\{f^{j_\ell}\}$, we may also suppose that f^{j_ℓ} converges normally to a holomorphic function $\widetilde{f}$. It follows that $\widetilde{f}'(P) = 1$. We claim, in fact, that $\widetilde{f}(z) \equiv z$.

To prove this assertion, suppose not. Assume for simplicity that $P = 0$. Then, for z near 0,

$$\widetilde{f}(z) = 0 + z + \text{(higher order terms)}.$$

Let $a_m z^m$, $m \geq 2$, be the first nonvanishing power of z which appears after z (since we are assuming that $\widetilde{f}$ is not the function z, there must be one). Notice that

$$\begin{aligned} \widetilde{f}^2 &\equiv \widetilde{f} \circ \widetilde{f} = z + 2a_m z^m + \ldots \\ \widetilde{f}^3 &\equiv \widetilde{f} \circ \widetilde{f} \circ \widetilde{f} = z + 3a_m z^m + \ldots \\ &\vdots \\ \widetilde{f}^k &\equiv \widetilde{f} \circ \widetilde{f} \circ \cdots \circ \widetilde{f} = z + k a_m z^m + \ldots. \end{aligned} \tag{2.19.1}$$

On the other hand, $\{\widetilde{f}^k\}$ forms a normal family, so there is a subsequence $\widetilde{f}^{k_q}$ converging normally on U. Hence

$$\left(\frac{\partial}{\partial z}\right)^m \widetilde{f}^{k_q}(P)$$

converges. But (2.19.1) clearly shows that

$$\left(\frac{\partial}{\partial z}\right)^m \widetilde{f}^k(P) = m! \cdot k \cdot a_m$$

blows up with k. This contradiction can only be resolved if $a_m = 0$. Therefore $\widetilde{f}(z) \equiv z$.

Now we have that $f^{j_\ell}(z) \to z$ normally. We claim that this implies that f is a conformal map. Indeed, consider the family $\{f^{j_\ell - 1}\}$. It is normal, so there is a subsequence, call it $\{f^{j_r - 1}\}$, which converges normally to a limit g. Notice that g is not constant because $g'(P)$ is not zero. Then

$$\begin{aligned} f \circ g(z) &= f \circ \lim_{r \to \infty} f^{j_r - 1}(z) \\ &= \lim_{r \to \infty} f^{j_r}(z) \\ &\equiv z. \end{aligned} \tag{2.19.2}$$

Similarly,

$$g \circ f(z) \equiv z.$$

Thus f is one-to-one and onto. It is certainly holomorphic, so f is a conformal map as claimed. □

It is instructive to view Theorem 2.19 as a generalization of the uniqueness part of the Schwarz lemma. For simplicity, take U to be a bounded domain. Then U is contained in some bounded disc, and that disc has nondegenerate Kobayashi metric. Therefore, by the distance decreasing property of the Kobayashi metric applied to the inclusion map, the domain U also has nondegenerate Kobayashi metric. Suppose that f is a holomorphic function mapping U to U that fixes a point P in U. The distance-decreasing property of the Kobayashi metric tells us that $|f'(P)| \leq 1$. Theorem 2.19 now says that equality obtains if and only if f is one-to-one and onto.

2.3 Completeness of the Carathéodory and Kobayashi Metrics

Capsule: Not surprisingly, it is important and useful to know that the invariant metric that we are studying is complete. With that information, we can safely perform the usual operations of analysis.

This section treats the completeness of the Carathéodory and Kobayashi metrics.

In this section, we prove that, if U is a domain with a reasonably nice boundary, then it is complete, as a metric space, when equipped with either the Carathéodory or Kobayashi metric.

We begin by noting that, on compact subsets of a bounded domain, all of our metrics are comparable:

PROPOSITION 2.23 *Let $U \subseteq \mathbb{C}$ be a bounded domain. Let $L \subseteq U$ be a compact subset. Let ρ_C, ρ_K, and ρ_E be the Carathéodory, Kobayashi, and Euclidean metrics on U, respectively. Then there are constants C_1, C_2, C_3, and C_4 such that, for $z \in L$,*

$$C_1 \leq \frac{\rho_C(z)}{\rho_E(z)} \leq C_2$$

and

$$C_3 \leq \frac{\rho_K(z)}{\rho_E(z)} \leq C_4.$$

[We stress that these four constants certainly depend both on L and on U.]

Proof: We will prove the second set of inequalities. The proof of the first inequalities is similar.

Now let P be a point in U. Let r be a positive number that is less than one third of the Euclidean distance of P to the boundary. Let $\overline{D}(P, r) \subseteq \overline{D}(P, 2r)$ be closed discs in U. If $z \in D(P, r)$ then

$$\rho_K^{D(P,2r)}(z) \geq \rho_K^U(z).$$

Here we use a superscript on ρ to indicate the domain with respect to which the metric is being computed.

On the other hand,

$$\rho_K^{D(P,2r)}(z) = \frac{1}{2r} \cdot \rho_K^{D(0,1)}((z-P)/[2r]).$$

And of course we know the Kobayashi metric on the unit disc explicitly. It equals the Poincaré metric. Hence $\rho_K^{D(0,1)}((z-P)/[2r]) \leq 4/3$. Putting our estimates together, we find that

$$\rho_K^U(z) \leq \frac{4}{3} \cdot \frac{1}{2r}.$$

This is one half of our estimate.

For the other half, notice that since U is bounded it is contained in some large disc open $D(0, R)$. Then, for $z \in D(P, r)$,

$$\rho_K^U(z) \geq \rho_K^{D(0,R)}(z).$$

Now, by conformal mapping, this last right-hand expression is equal to $[1/R]\rho_K^{D(0,1)}(z/[R])$. Thus it is at least equal to $1/R$.

In summary, we have proved, for points $z \in D(P, r)$, that

$$\frac{4}{6r} \geq \rho_K^U(z) \geq \frac{1}{R}.$$

The compact set $L \subset U$ can be covered by finitely many discs of the form $D(P, r)$. So, in the end, we obtain a uniform estimate

$$C_3 \leq \rho_K^U(z) \leq C_4$$

for all $z \in L$. However the Euclidean metric is constantly equal to 1. So this last line just says

$$C_3 \rho_E(z) \leq \rho_K^U(z) \leq C_4 \rho_E(z),$$

and that is what we wished to prove. □

COROLLARY 2.24 *Let $U \subseteq \mathbb{C}$ be a bounded domain. The topology induced by any of the Carathéodory, Kobayashi, or Euclidean metrics is equivalent to the other two.*

Proof: Since the metrics are comparable, the induced balls are comparable. The balls form a sub-basis for the topology. So the topologies are comparable. □

Now we turn to the treatment of completeness of our two new metrics on a reasonable class of domains. It is enough to prove completeness for the Carathéodory metric, since any sequence which is Cauchy in the Kobayashi metric is also Cauchy in the Carathéodory metric. We begin by defining what we mean by "reasonably nice boundary."

Definition 2.25 A curve $\gamma : [a, b] \to \mathbb{C}$ is called a *closed, twice continuously differentiable curve* if γ is twice continuously differentiable (with one-sided derivatives as usual at the endpoints), if $\dot{\gamma}$ is never 0, if $\gamma(a) = \gamma(b)$, and if the one-sided derivatives of γ, up to and including order 2, at a equal those at b. We write "γ is C^2 and closed." More generally, we may define γ to be *closed, k times continuously differentiable* (for $k \geq 1$) if γ is k times continuously differentiable, (with one-sided derivatives as usual at the endpoints), if $\dot{\gamma}$ is never 0, if $\gamma(a) = \gamma(b)$, and if the one sided derivatives of γ, up to and including order k, at a equal those at b. We write "γ is C^k and closed."

Definition 2.26 A domain is said to have C^k *boundary* if the boundary consists of finitely many C^k, closed curves.

In case a function is C^k for every k, or a boundary is C^k for every k, then we say that the function or boundary is C^∞ or infinitely differentiable.

In geometric analysis, an alternative definition of C^k boundary often proves to be convenient. We think of a domain U with C^k (resp. C^∞) boundary as one which can be specified by

$$U = \{z \in \mathbb{C} : \rho(z) < 0\},$$

with ρ a C^k (resp. C^∞) function such that $\nabla \rho \neq 0$ on ∂U. We call ρ a *defining function* for U. For example, the disc may be specified in this way as

$$D = \{z \in \mathbb{C} : \rho(z) = |z|^2 - 1 < 0\}.$$

A bounded domain with infinitely many holes will *not* have C^k boundary, $k \geq 1$, since any defining function will fail to be smooth at an accumulation point of the holes. See [KRA2] for a thorough treatment of defining functions.

Example 2.27 The curve $\gamma(t) = e^{it}$, $0 \leq t \leq 2\pi$, is a closed, twice continuously differentiable curve (indeed, it is C^k for every k, or infinitely differentiable). See Figure 2.1. Notice that the agreement of the first and second derivatives at the endpoints causes a smooth transition in the figure at the point $1 + i0$ where the two ends meet.

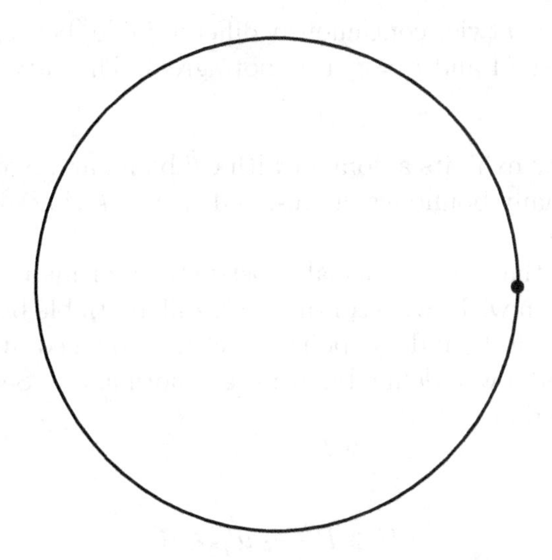

Figure 2.1: A closed, twice continuously differentiable curve.

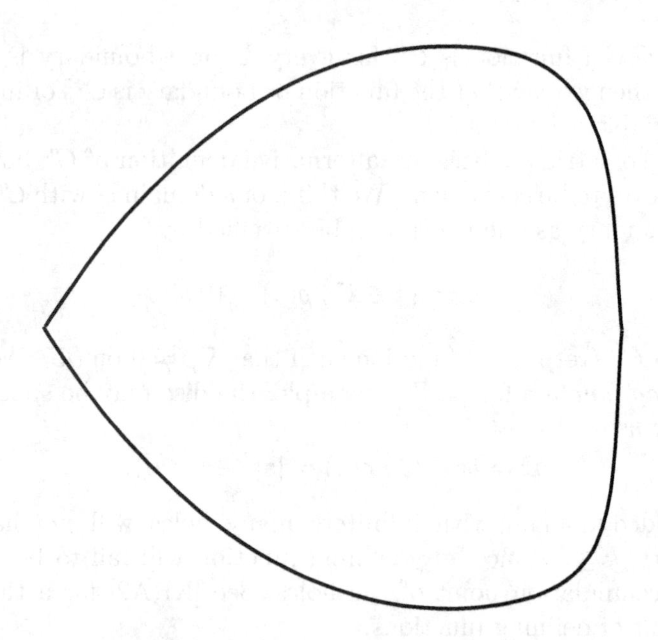

Figure 2.2: A boundary curve that is *not* twice continuously differentiable.

The curve

$$\mu(t) = (\cos 2t)(\cos t) + i(\cos 2t)(\sin t), \qquad -\frac{\pi}{4} \le t \le \frac{\pi}{4}$$

is closed, but *not* twice continuously differentiable (because the derivatives at $t = -\pi/4$ and $t = \pi/4$ do not agree). The curve is shown in Figure 2.2.

Figure 2.3 exhibits a domain with C^k boundary. Note that there are finitely many boundary curves, and each is k times continuously differentiable.

Observe that, by the usual constructions of multi-variable calculus, a domain with twice continuously differentiable boundary possesses at each boundary point P a well defined unit outward normal ν_P and a well defined unit inward normal ${\nu'}_P$. See Figure 2.4.

The functions

$$\partial U \ni P \longrightarrow \nu_P \in \mathbb{R}^2$$

and

$$\partial U \ni P \longrightarrow {\nu'}_P \in \mathbb{R}^2$$

are then continuously differentiable.

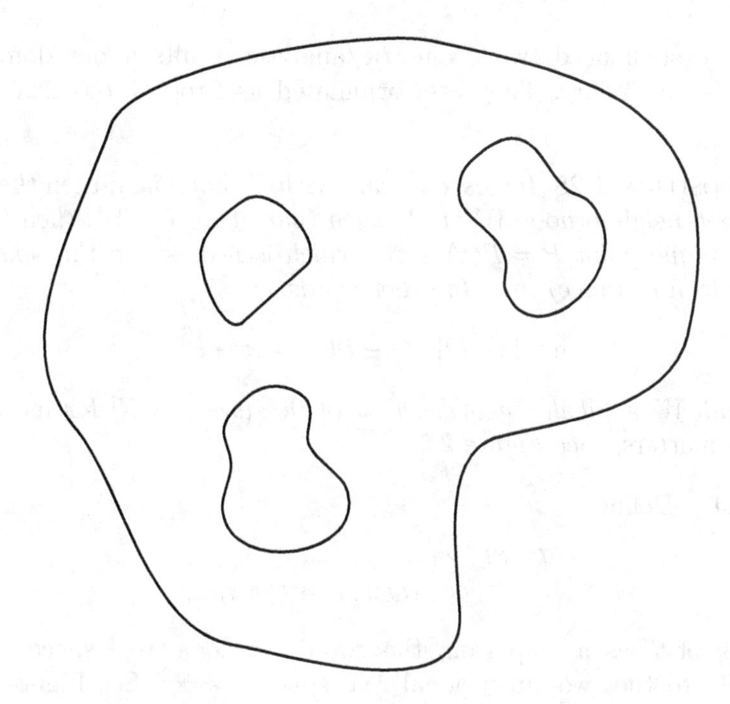

Figure 2.3: A domain with C^k boundary.

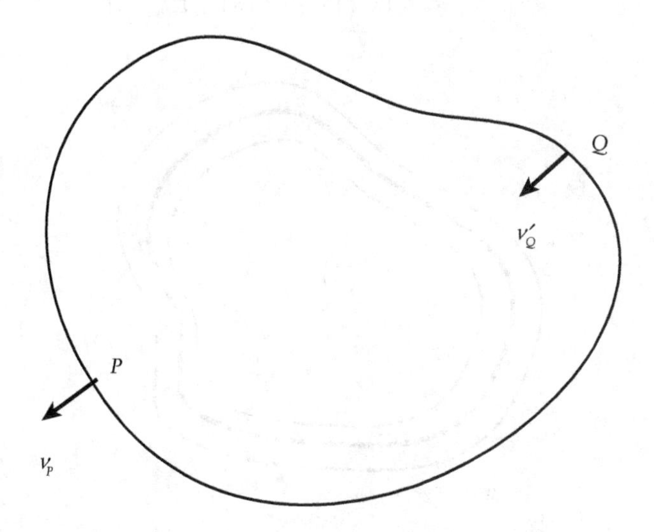

Figure 2.4: The unit inward and unit outward normals.

We shall need two geometric/analytic results about domains with C^2 boundary. They are formulated as Propositions 2.28 and 2.30.

PROPOSITION 2.28 *If U is a domain with C^2 boundary. then there is an open neighborhood W of ∂U such that, if $z \in U \cap W$, then there is a unique point $P = P(z) \in \partial U$ which is nearest (in the sense of Euclidean distance) to z. In other words,*

$$\inf\{|z - Q| : Q \in \partial U\} = |z - P|.$$

We call W a tubular neighborhood of ∂U (see [MUN] for more on these matters). See Figure 2.5.

Proof: Define

$$\begin{aligned} T : \partial U \times (-1,1) &\longrightarrow \mathbb{C} \\ (Q,t) &\longmapsto Q + t\nu'_P. \end{aligned}$$

Think of T as a map from the two-dimensional real space $\partial U \times (-1,1)$ to the two-dimensional real space $\mathbb{C} \approx \mathbb{R}^2$. See Figure 2.6. (In fact, the reader may find it convenient to consider this alternative definition of T. Let $\gamma : [0,1] \to \mathbb{C}$ be a C^2 parametrization of ∂U. Then set

$$\begin{aligned} T : [0,1] \times (-1,1) &\longrightarrow \mathbb{C} \\ (s,t) &\longmapsto \gamma(s) + t\nu'_{\gamma(s)}.) \end{aligned}$$

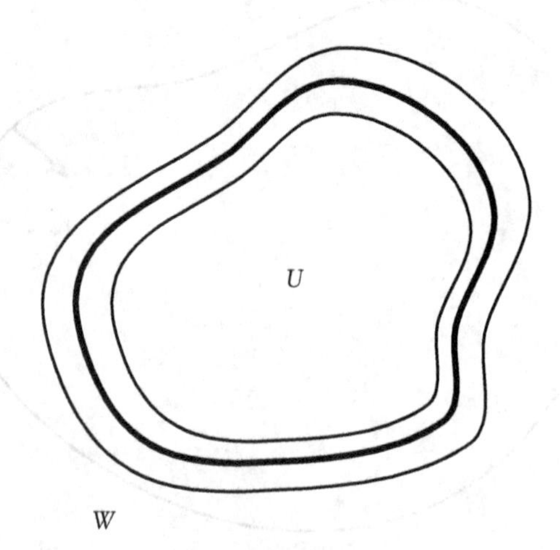

Figure 2.5: A tubular neighborhood of the boundary.

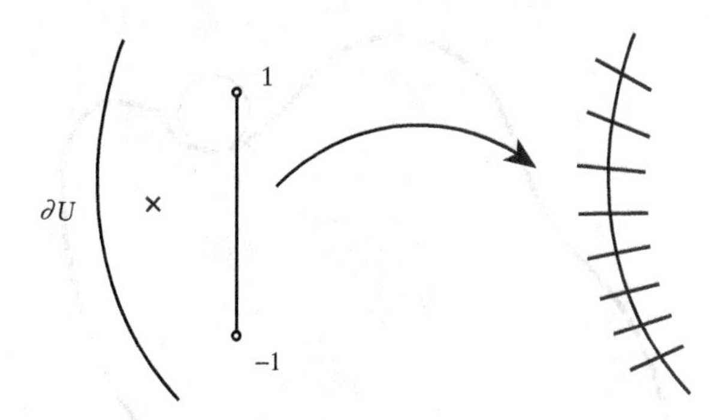

Figure 2.6: Construction of a tubular neighborhood.

Fix a point $Q_0 \in \partial U$. Assume, without loss of generality, that the tangent line to ∂U at Q_0 is horizontal (this may be achieved by rotating the coordinate system). We may also normalize the tangent vector to be of unit length.

Then the Jacobian matrix of T at $(Q_0, 0)$ is

$$\begin{pmatrix} 1 & 0 \\ 0 & 1 \end{pmatrix}$$

which is invertible. By the inverse function theorem (see [RUD1], [MUN], or [KRP1]), there is a neighborhood V_0 of $(Q_0, 0)$ in $\partial U \times (-1, 1)$ on which T is invertible. Then $T(V_0) \equiv W_{Q_0}$ is a neighborhood of Q_0 with the property that if $z \in W_{Q_0}$, then there is a unique point $Q \in \partial U$ and a unique $t \in (-1, 1)$ such that

$$z = Q + t\nu'_Q.$$

It follows that Q is (locally) the unique nearest point in ∂U to z, and the distance of z to Q is $|t|$. We set

$$W = \bigcup_{Q_0 \in \partial U} W_{Q_0}$$

and we are finished. □

REMARK 2.29 The hypothesis of C^2 boundary was needed in order to apply the inverse function theorem.. Explain why. The analogous result is false as soon as the boundary smoothness is less than C^2 (see [KRP2]).

PROPOSITION 2.30 *Let $U \subseteq \mathbb{C}$ be a bounded domain with C^2 boundary. Then there is an $r_0 > 0$ such that, for each $P \in \partial U$, there is a*

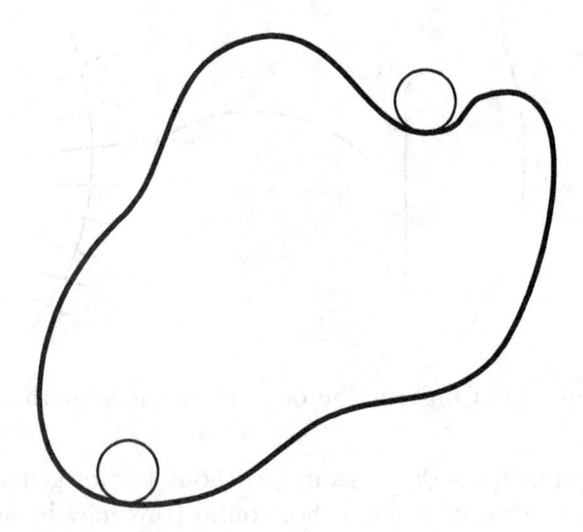

Figure 2.7: An internally tangent and an externally tangent disc.

disc $D(C(P), r_0)$ of radius r_0 which is externally tangent to ∂U at P. There is also a disc $D(C'(P), r_0)$ which is internally tangent to ∂U at P. These discs have the further property that $\overline{D}(C(P), r_0) \cap \partial U = \{P\}$ and $\overline{D}(C'(P), r_0) \cap \partial U = \{P\}$.

REMARK 2.31 Refer to Figure 2.7 and notice that a disc $\mathbf{d}$ is internally tangent to ∂U at P if $\mathbf{d} \subseteq U$, $\overline{\mathbf{d}} \ni P$, and $\partial\mathbf{d}$ and ∂U have the same tangent line at P.

A similar remark applies to externally tangent discs.

Proof of Proposition 2.30: This follows from basic calculus: To each point $P \in \partial U$ there corresponds a radius of curvature $r(P)$ and a center of curvature $c(P)$ (here we are using the word "curvature" in the classical Euclidean sense). The point $c(P)$ is $r(P)$ units along the *principal normal* from P.

The corresponding circle is called the circle of curvature. See [BLK] for details. If the circle of curvature is *inside* the domain, take $C'(P) = c(P)$, and take $C(P)$ to be the reflected point in ∂U. See Figure 2.8. If the circle of curvature is *outside* the domain, take $C(P) = c(P)$, and take $C'(P)$ to be a reflected point in ∂U.

Now, referring to [BLK], we see that $r(P)$ depends on the second derivative of the boundary curve γ. In particular, $r(P)$ is a continuous function of P. We take r_0 to be the (positive) minimum of r. We would like to think that we are finished.

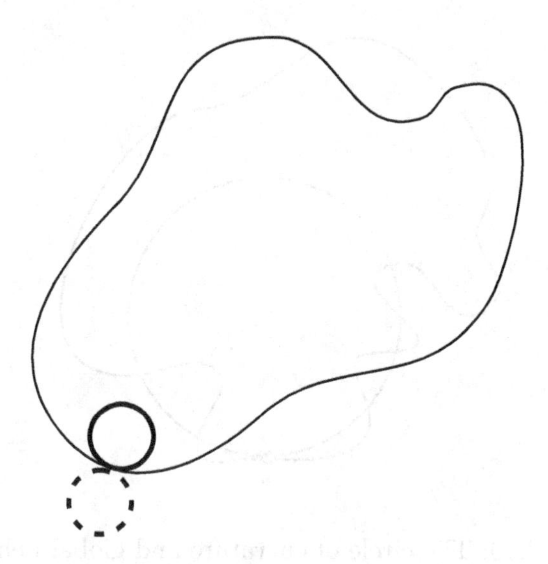

Figure 2.8: The circle of curvature.

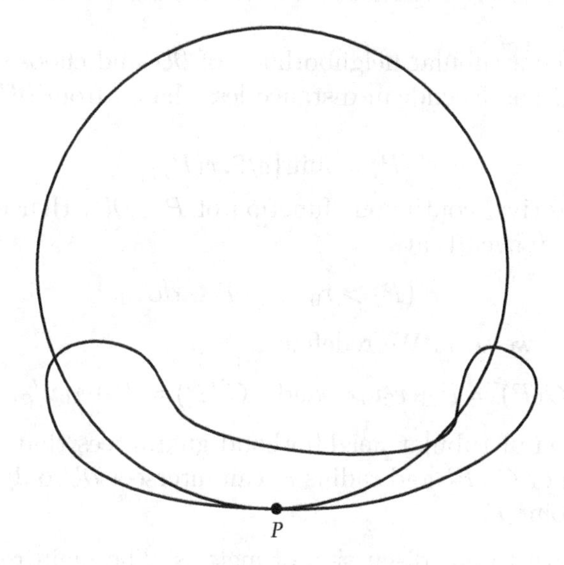

Figure 2.9: The circle of curvature and global behavior.

Unfortunately, in spite of our best intentions, we may have a situation as in Figures 2.9 or 2.10: The circle of curvature has the right behavior *at* P, but it does not take into account the global behavior of U. As a result, it may be neither completely internal

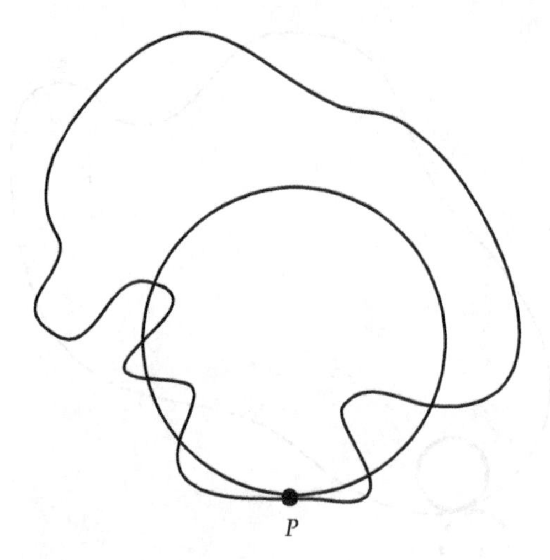

Figure 2.10: The circle of curvature and global behavior.

to U nor completely external to U. We therefore make the following adjustment.

Let W be a tubular neighborhood of ∂U and choose $\epsilon > 0$ such that if $z \in \mathbb{C}$ has Euclidean distance less than ϵ from ∂U, then z is in W. Define

$$r^*(P) = \min\{\epsilon/2, r(P)\}.$$

This is a positive, continuous function of $P \in \partial U$. Hence there is a number $r_0 > 0$ such that

$$r^*(P) > r_0 \qquad \forall P \in \partial U.$$

This is the r_0 we seek. We redefine

$$C(P) = P + r_0\nu_P \quad \text{and} \quad C'(P) = P + r_0{\nu'}_P.$$

The definition of tubular neighborhood guarantees that a disc with center $C(P)$ or $C'(P)$ and radius r_0 can intersect ∂U only at the one boundary point P. □

We return to our discussion of metrics. The main result of this section is the following.

Prelude: This is one of our fundamental completeness results. Obviously it is necessary if we are going to do any analysis.

Since the Kobayashi metric dominates the Carathéodory metric, we may immediately conclude that the Kobayashi metric is also complete.

THEOREM 2.32 *If $U \subseteq \mathbb{C}$ is a bounded domain with C^2 boundary, then U is complete in the Carathéodory metric.*

COROLLARY 2.33 *The domain U is also complete in the Kobayashi metric.*

Proof of the Theorem: Let $z \in U$ be in a tubular neighborhood W of ∂U, as provided by Proposition 2.28. Let P be the nearest boundary point to z and $D(C(P), r_0)$ the externally tangent disc provided by Proposition 2.30. The map

$$\begin{aligned} \mathbf{i}_P : U &\longrightarrow D(C(P), r_0) \\ \zeta &\longmapsto C(P) + \frac{(r_0)^2}{\zeta - C(P)} \end{aligned}$$

is holomorphic and inverts U into the disc $D(C(P), r_0)$. The map

$$\begin{aligned} \mathbf{j}_P : D(C(P), r_0) &\longrightarrow D(0,1) \\ \zeta &\longmapsto \frac{\zeta - C(P)}{r_0} \end{aligned}$$

is holomorphic. To estimate the Carathéodory metric at z, we use the distance-decreasing property of the Carathéodory metric:

$$\begin{aligned} F_C^U(z) &\geq ((\mathbf{j}_P \circ \mathbf{i}_P)^* F_C^{D(0,1)})(z) \\ &\equiv |(\mathbf{j}_P \circ \mathbf{i}_P)'(z)| F_C^{D(0,1)}(\mathbf{j}_P(\mathbf{i}_P(z))). \end{aligned} \tag{2.32.1}$$

We now estimate the various terms in this last expression.

First,

$$\begin{aligned} |(\mathbf{j}_P \circ \mathbf{i}_P)'(z)| &= \left| \frac{1}{r_0} \cdot (\mathbf{i}_P)'(z) \right| \\ &= \left| \frac{1}{r_0} \cdot \frac{(r_0)^2}{(z - C(P))^2} \right| \\ &= \left| \frac{r_0}{(z - C(P))^2} \right| \\ &= \left| \frac{r_0}{(\delta + r_0)^2} \right|, \end{aligned} \tag{2.32.2}$$

where δ is the Euclidean distance of z to P. See Figure 2.11.

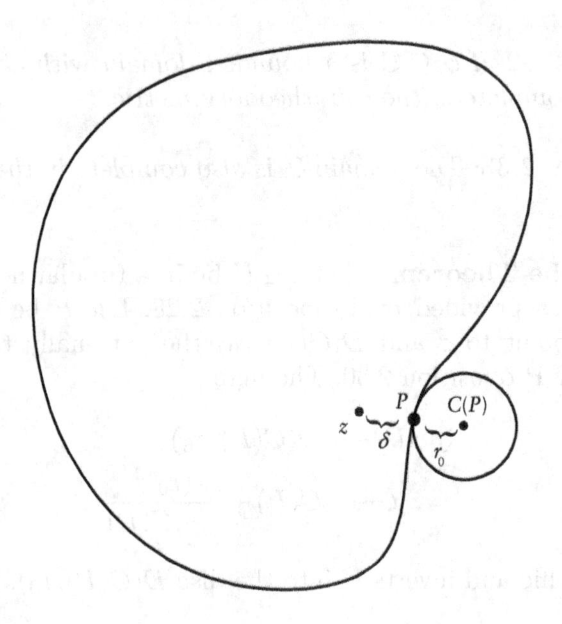

Figure 2.11: Estimating the Carathéodory metric.

Also

$$\mathbf{j}_P \circ \mathbf{i}_P(z) = \mathbf{j}_P\left(\frac{(r_0)^2}{z - C(P)} + C(P)\right)$$
$$= \frac{r_0}{z - C(P)};$$

hence

$$|\mathbf{j}_P \circ \mathbf{i}_P(z)| = \frac{r_0}{\delta + r_0} = 1 - \frac{\delta}{r_0 + \delta}.$$

Recall that, on the unit disc, the Carathéodory metric coincides with the Poincaré metric. It follows from our calculation of the Poincaré metric that

$$\begin{aligned} F_C^{D(0,1)}(\mathbf{j}_P \circ \mathbf{i}_P(z)) &= \frac{1}{(\delta/(r_0+\delta))(2 - \delta/(r_0+\delta))} \\ &\geq \frac{r_0 + \delta}{2\delta} \\ &\geq \frac{r_0}{2} \cdot \frac{1}{\delta}. \end{aligned} \tag{2.32.3}$$

In summary, using (2.32.1)), (2.32.2), and (2.32.3), we have

$$F_C^U(z) \geq \frac{r_0}{(\delta + r_0)^2} \cdot \frac{r_0}{2} \cdot \frac{1}{\delta} \geq C_0 \cdot \frac{1}{\delta},$$

where C_0 is a positive constant which depends only on r_0.

But this is precisely the estimate which enables us, in Section 5.2.4 below, to prove that the disc is complete in the Poincaré metric. We leave it now as an exercise to provide the details which show that the Carathéodory distance from any fixed interior point $P_0 \in U$ to a point with Euclidean distance δ from the boundary has size $C \cdot (1 + |\log 1/\delta|)$ and to conclude that U is complete in the Carathéodory metric. □

Exercise: Exploit the internally tangent disc at each boundary point to prove that there is a constant C_1 such that

$$F_C^U(z) \leq C_1 \cdot \frac{1}{\delta}.$$

(Hint: Use the inclusion map from the internal disc to the domain, together with the distance-decreasing property of the Carathéodory metric.)

Exercise: Let

$$U = D \setminus \{0\}$$

be the punctured disc. See Figure 2.12. This domain does *not* have C^2 boundary. Use the Riemann removable singularities theorem and Cauchy estimates to determine the behavior of the Carathéodory metric near the boundary point 0 of U. Conclude that U is not complete in the Carathéodory metric. What can you say about the Kobayashi metric?

COROLLARY 2.34 *Let U be any bounded, finitely connected domain in $\mathbb{C}$ (that is, the complement of U has finitely many connected components). Further assume that each boundary component of ∂U is a Jordan curve. Then the Carathéodory metric on U is complete.*

Proof: Since completeness of the Carathéodory metric is a conformal invariant, it is enough to show that U is conformally equivalent to a domain $\widetilde{U}$ having the property that $\partial\widetilde{U}$ has C^2 boundary.

To achieve this end, imagine a planar domain with a single hole. Assume that the outer boundary is a Jordan curve, and the boundary of the hole is also a Jordan curve. Fill in the hole to obtain a simply

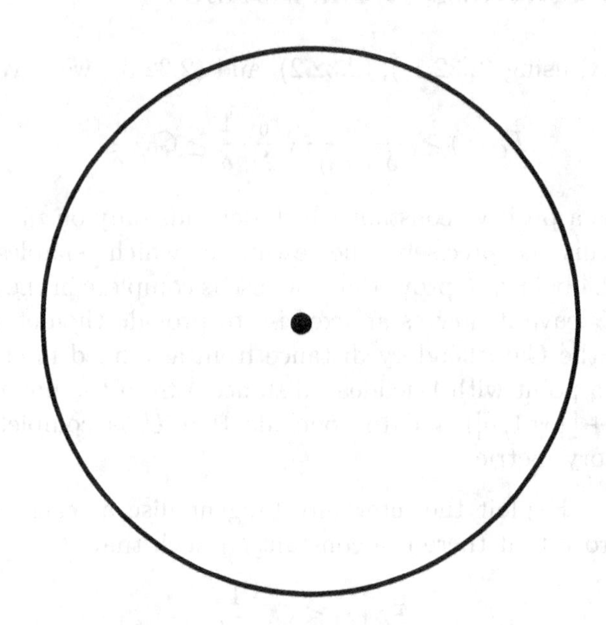

Figure 2.12: The punctured disc.

connected domain U', and use the Riemann Mapping theorem to map U' to the disc. Now unfill the hole. See Figure 2.13. We now have a domain U'' with a circle as the outer boundary and a Jordan curve as the inner boundary.

Let A be the bounded component of the complement of U'' and let p be a point in the interior of A. Perform an inversion in p (i.e., $z \mapsto 1/(z-p)$). This maps ∂A to the outer boundary of a new domain, and the former (outer) boundary of U'' to the boundary of an interior hole (again see Figure 2.13). Now fill in the new hole, and use the Riemann Mapping theorem to map the (filled in) new domain to a disc. Finally, perform another inversion (Figure 2.13). Thus the original hole (that is, A) has been converted to a smooth, circular hole.

If ∂U has k components then k successive applications of inversion and the Riemann Mapping theorem, as above, give rise to a conformal mapping of U to a domain consisting of the unit disc with $k-1$ smoothly bounded domains removed. That completes the proof. See [AHL2] for further details. □

We now use the completeness of the Carathéodory metric to prove a version of the Lindelöf principle using the geometric point of view. First we need some terminology.

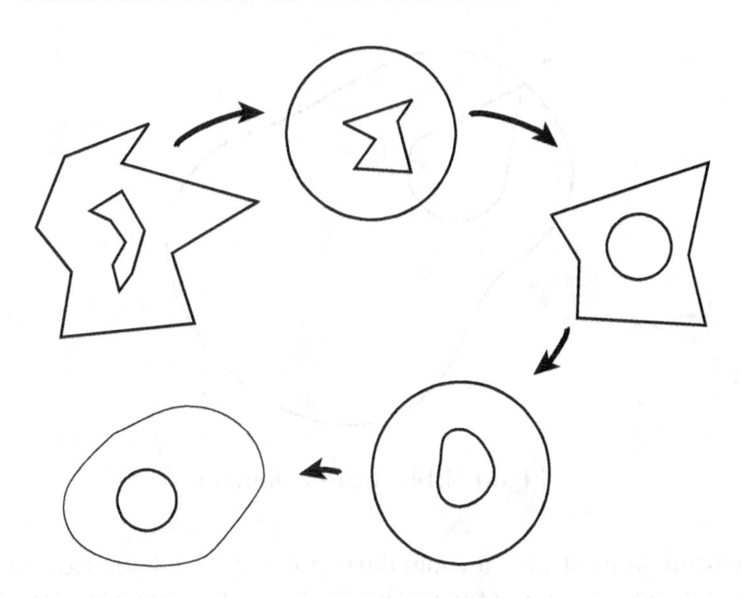

Figure 2.13: Another inversion.

Let U be a domain with C^2 boundary. Choose a point $P \in \partial U$ and let ν_P be the unit outward normal. If f is a continuous function on U, we say that f has *radial boundary limit* ℓ at P provided that

$$\lim_{r \to 0^+} f(P - r\nu_P) = \ell.$$

If $\alpha > 1$, set

$$\Gamma_\alpha(P) = \{z \in U : |z - P| < \alpha \cdot \text{dist}(z, \partial U)\}.$$

Here the expression $\text{dist}(z, \partial U)$ denotes Euclidean distance of z to ∂U. We call the domain Γ_α a *Stolz domain* or *nontangential approach domain* at the point P. See Figure 2.14.

We say that the function f has *nontangential limit* ℓ at P if, for each $\alpha > 1$, we have

$$\lim_{\Gamma_\alpha(P) \ni z \to P} f(z) = \ell.$$

Clearly the possession of a nontangential limit at P is a stronger condition than the possession of a radial limit at P. For example, on the unit disc, the continuous function

$$f(z) = \frac{y}{1 - x}$$

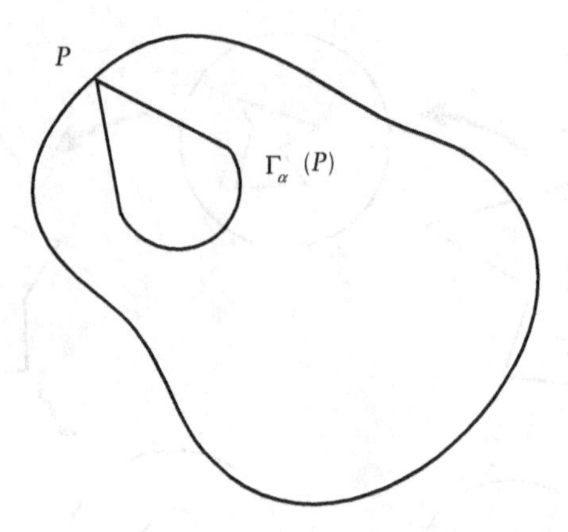

Figure 2.14: A Stolz domain.

has radial limit 0 at the boundary point $P = 1 + i0$. However, it does not possess a nontangential limit at P (exercise—try $P_j = (1 - 1/j) + i/j$ for $j = 1, 2, \ldots$). It is therefore surprising that the Lindelöf principle tells us that, for bounded, holomorphic functions, the two notions of boundary limit coincide:

Prelude: This result is a version of the Lindelöf principle, formulated in geometric language. It is also very natural to conceive the proof in analytic argot.

THEOREM 2.35 *Let U_1, U_2 be bounded domains with C^2 boundary and let*

$$f : U_1 \longrightarrow U_2$$

be holomorphic. If $P \in \partial U_1$, $Q \in \partial U_2$, and f has radial limit Q at P then f has nontangential limit Q at P.

Proof: If z is an element of one of our domains U_j and if $s > 0$ then we let $B(z, s)$ denote the metric ball with center z and radius s in the Carathéodory metric for U_j. For $P \in \partial U_1$, $r_0 > 0$, and $\beta > 0$ fixed, we define

$$\mathcal{M}_\beta(P) = \bigcup_{0<r<r_0} B(P - r\nu_P, \beta).$$

The estimate

$$F_C^U(z) \approx \frac{C}{\text{dist}(z, \partial U)} \tag{2.35.1}$$

makes it a tedious, but not difficult, exercise to calculate that the domains $\mathcal{M}_\beta$ are comparable to the domains Γ_α (in this last formula, and in what follows, "dist" means Euclidean distance). In point of fact, suppose that z lies in some $\Gamma_\alpha(p)$, some $p \in \partial U_j$. Let us denote by τ_p the inward normal segment emanating from that boundary point p. Using the estimate (2.35.1), one can then estimate that $\mathrm{dist}_C(z, \tau_p) \leq C \cdot \alpha$. For the converse estimate, assume that $z \notin \Gamma_\alpha(p)$ and the same estimate shows that $\mathrm{dist}_C(z, \tau_p) \geq C \cdot \alpha$.

Thus we see that

$$\lim_{\Gamma_\alpha(P)\ni z\in P} f(z) = \ell, \qquad \forall \alpha > 1$$

iff

$$\lim_{\mathcal{M}_\beta(P)\ni z\to P} f(z) = \ell, \qquad \forall \beta > 0. \tag{2.35.2}$$

Thus it is enough to prove (2.35.2).

Since

$$\mathcal{M}_\beta(P) = \bigcup_{0<r<r_0} B(P - r\nu_P, \beta),$$

the distance-decreasing property of f with respect to the Carathéodory metric implies that

$$f(\mathcal{M}_\beta(P)) \subseteq \bigcup_{0<r<r_0} B(f(P - r\nu_P), \beta).$$

Pick $\epsilon > 0$. By the radial limit hypothesis, there is a $\delta > 0$ such that, if $0 < t < \delta$, then

$$|f(P - t\nu_P) - Q| < \epsilon.$$

For such a t, if $z \in B(P - t\nu_P, \beta)$ then

$$f(z) \in B(f(P - t\nu_P), \beta).$$

But

$$\mathrm{dist}(f(P - t\nu_P), \partial U_2) \leq \mathrm{dist}(f(P - t\nu_P), Q) < \epsilon.$$

Therefore the estimate (2.35.1) implies that the ball $B(f(P-t\nu_P),\beta)$ has Euclidean radius not exceeding $C\cdot\epsilon$. Here C depends on β, but β has been fixed once and for all. Thus

$$|f(z)-f(P-t\nu_P)|<C\epsilon, \qquad \forall z\in B(P-t\nu_P,\beta).$$

We conclude that

$$\begin{aligned}|f(z)-Q| &\leq |f(z)-f(P-t\nu_P)|+|f(P-t\nu_P)-Q| \\ &\leq C\epsilon+\epsilon=C'\epsilon.\end{aligned}$$

This is the desired conclusion. □

It is not difficult to see that nontangential approach is the broadest possible method for calculating boundary limits of bounded, holomorphic functions. See [PRI] and [DIK] for explicit constructions of examples to show that the nontangential approach is the sharp condition. These references also contains a complete account of the positive results, as well as the history, of the theory of boundary behavior of holomorphic functions.

Chapter 3

Normal Families

Prologue: Montel's notion of normal family is a simple, elegant ideas that has unexpected power and beauty. The proof of Montel's theorem uses key ideas from real variable theory, but the end result is something that can only be true for holomorphic functions.

Normal families are used in the modern proof of the Riemann Mapping theorem—indeed, they provide a nice correction to the error in Riemann's original proof. Also, they are a key part of the theory of automorphism groups, they are used in partial differential equations, and they are a decisive tool in classical function theory.

The modern way to think about Montel's theorem is that it is a compactness statement: A family of functions that satisfies a certain boundedness property must have a convergent subsequence. Viewed in this manner, the idea of normal families fits naturally into many parts of mathematical analysis, from functional analysis to hard analysis.

One of the profound new ideas to grow out of normal families is Lehto and Virtanen's concept of normal function. This is one of the exciting new classes of functions that is being studied even today.

In this chapter, we give a thorough grounding in the theory of normal families, and provide some pointers to further reading.

3.1 Montel's Theorem

Capsule: The idea of normal family is one of the central and profound ideas of complex function theory. Montel's theorem is the pivotal result in the subject, and it plays a central role in the subject even today.

The proof of Montel's theorem is an interesting blend of real and complex analysis. But the final result is pure function theory.

Ideas related to normal families continue to be developed even today, and the applications of normal families are broad and diffuse.

The results of the previous chapter have already suggested that there is special interest in holomorphic functions that satisfy some extremal property. The special nature of such extremal problems is this: For any $\epsilon > 0$, there will be a holomorphic function f_ϵ that comes within ϵ of being the extremum (e.g., the least upper bound in our special case from the previous chapters). Thus there is a sequence f_j such that the f_j come ever closer to being the extremum; *but one would like to assert that (some subsequence of) the* f_j *actually converge to an "extremal function"* f_0.

As noted elsewhere, what we are in effect looking for here is a compactness result: If a collection of functions is bounded in some topology, then it converges in some other topology. That turns out to be a useful way to think about normal families, and about Montel's theorem.

We now develop a versatile set of tools that allows us to extract a convergent subsequence from a fairly general family of holomorphic functions. We begin with a definition that will be useful in the formulation of our main result.

Definition 3.1 A sequence of functions f_j on an open set $U \subseteq \mathbb{C}$ is said to *converge normally* to a limit function f_0 on U if $\{f_j\}$ converges to f_0 uniformly on compact subsets of U. That is, the convergence is normal if, for each compact set $K \subseteq U$ and each $\epsilon > 0$, there is a $J > 0$ (depending on K and ϵ) such that, when $j > J$ and $z \in K$, then $|f_j(z) - f_0(z)| < \epsilon$.

With U and $\{f_j\}$ as above, we say that the sequence $\{f_j\}$ is *compactly divergent* if, for each pair of compact sets $K \subseteq U$ and $L \subseteq \mathbb{C}$, there is an $N > 0$ such that if $j > N$, then $f_j(z) \notin L$ for all $z \in K$.

The correct definition of normal family, in view of these definitions, is next.

Definition 3.2 Let $U \subseteq \mathbb{C}$ be a domain. Let $\mathcal{F} = \{f_\alpha\}_{\alpha \in \mathcal{A}}$ be a family of functions on U. We say that $\mathcal{F}$ is a *normal family* if each subsequence $\{f_j\} \subseteq \mathcal{F}$ either has a subsequence $\{f_{j_k}\}$ that converges normally or else has a subsequence $\{f_{j_\ell}\} \subseteq \mathcal{F}$ that is compactly divergent.

The classical definition of normal family only treated the case where there was a normally convergent subsequence of functions. The idea of compactly divergent family is a relatively newer idea. But it is very natural from a geometric point of view. It means that we are thinking about functions taking values in the Riemann sphere. In other words, our new definition of normal family allows us to think about normal families of *meromorphic* functions.

Example 3.3 The functions $f_j(z) = z^j$ converge normally on the unit disc D to the function $f_0(z) \equiv 0$. The sequence does *not* converge uniformly on all of D to f_0, but it does converge uniformly on each compact subset of D.

Example 3.4 Let $U = \{z \in \mathbb{C} : 1/2 < |z| < 2\}$. Let $f_j(z) = z^j$ for $j = 1, 2, \ldots$ and $\mathcal{F} = \{f_j\}$. Then the family $\mathcal{F}$ is not normal on U. The functions are normally convergent on $\{|z| < 1\}$ and compactly divergent on $\{|z| > 1\}$. But they are neither on $\{|z| = 1\}$.

Prelude: This is the most basic and fundamental version of Montel's theorem. We shall see below that there are many ways to think about this result, some of them quite a bit more general.

THEOREM 3.5 (MONTEL'S THEOREM, FIRST VERSION) *Consider the family $\mathcal{F} = \{f_\alpha\}_{\alpha \in A}$ of holomorphic functions on a domain $U \subseteq \mathbb{C}$. Suppose that there is a constant $M > 0$ such that, for all $z \in U$ and all $f_\alpha \in \mathcal{F}$,*

$$|f_\alpha(z)| \leq M.$$

Then, for every sequence $\{f_j\} \subseteq \mathcal{F}$, there is a subsequence $\{f_{j_k}\}$ that converges normally on U to a limit (holomorphic) function f_0.

Montel's theorem is a remarkable compactness statement. For contrast in the real-variable situation, note that the functions $\{\sin kx\}_{k=1}^{\infty}$ are all bounded by the constant 1 on the interval $[0, 2\pi]$, yet there is no convergent subsequence—not even a pointwise convergent subsequence, much less a normally convergent one.

The proof of Montel's theorem uses powerful machinery from real analysis, such as the Ascoli-Arzela theorem. We refer to [GRK1] for a thorough treatment.

Proof of Montel's theorem Fix a compact set $K \subseteq U$. Choose a slightly larger compact set $L \subseteq U$ with its interior $\overset{\circ}{L}$ containing K. We may suppose that, for some $\eta > 0$, any two points $z, w \in K$ with $|z - w| < \eta$ have the property that the line segment connecting them lies in L. See Figure 3.1. Since L is compact, there is a number $r > 0$ such that if $\ell \in L$, then $\overline{D}(\ell, r) \subseteq U$. But then, for any $f \in \mathcal{F}$, we have from the Cauchy estimates that

$$|f'(\ell)| \leq \frac{M}{r} \equiv C_1.$$

Observe that this inequality is uniform in both ℓ and f. Now let $z, w \in K$. Fix an $f \in \mathcal{F}$. If $|z - w| < \eta$ and if $\gamma : [0, 1] \to \mathbb{C}$ is a parametrization of the line segment connecting z to w, then we have

$$\begin{aligned}
|f(z) - f(w)| &= |f(\gamma(1)) - f(\gamma(0))| \\
&= \left| \int_0^1 f'(\gamma(t)) \cdot \gamma'(t) dt \right| \\
&\leq C_1 \int_0^1 |\gamma'(t)| dt \\
&= C_1 |z - w|.
\end{aligned}$$

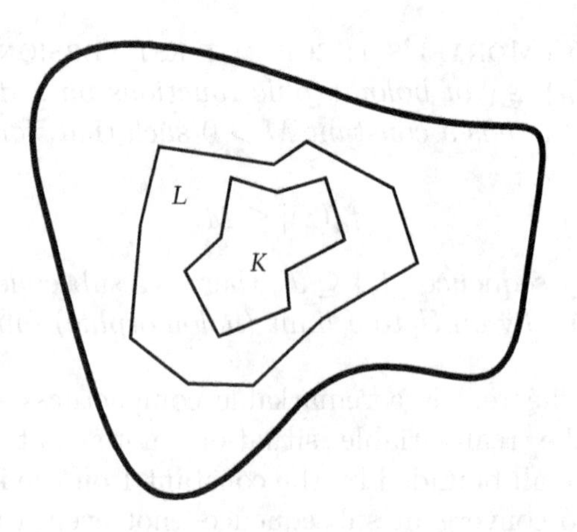

Figure 3.1: A compact set K in U.

Thus, for any $z, w \in K$ and any $f \in \mathcal{F}$ we have that, if $|z-w| < \eta$,

$$|f(z) - f(w)| \leq C|z - w|.$$

In particular, $\mathcal{F}$ is an equicontinuous family. The Ascoli-Arzelà theorem thus applies to show that any sequence $\{f_j\} \subseteq \mathcal{F}$ has a subsequence $\{f_{j_k}\}$ that converges uniformly on K.

Now we need an extra argument to see that a subsequence can be found that converges uniformly on every compact subset of U (the *same* sequence has to work for every compact set). Thus let $K_1 \subseteq K_2 \subseteq \cdots$ be compact sets such that $K_j \subseteq \overset{\circ}{K}_{j+1}$ for each j and $\bigcup_j K_j = U$ (e.g., we could let

$$K_j = \{z \in U : \text{distance}\,(z, \mathbb{C} \setminus U) \geq 1/j\} \cap \overline{D}(0, j)).$$

[In case $U = \mathbb{C}$, letting $K_j = \overline{D}(0, j)$ would do.] Then, by the argument we have just given, there is a sequence $\{f_j\} \subseteq \mathcal{F}$ such that $\{f_j\}$ converges uniformly on K_1. Applying the argument again, we have a subsequence of this sequence—call it $\{f_{j_k}\}$—that converges uniformly on K_2. Continuing in this fashion, we may find, for each index m, a subsequence of the $(m-1)$th subsequence that converges uniformly on K_m. Now we form a final sequence by setting

$$\begin{aligned} g_1 &= f_1, \\ g_2 &= f_{j_2}, \\ g_3 &= f_{j_{k_3}}, \\ &\text{etc.} \end{aligned}$$

Then the sequence $\{g_j\}$ is a subset of $\mathcal{F}$ and is a subsequence of each of the sequences that we formed above. So $\{g_j\}$ converges uniformly on each K_j. Now *any* compact subset L of U must be contained in some K_j since $\bigcup \overset{\circ}{K}_j$ is an open cover of L, which necessarily has a finite subcover. Therefore the sequence g_j converges uniformly on any compact subset of U. □

REMARK 3.6 Note that it would be incorrect to hope that we can find a subsequence $\{f_{j_k}\}$ that converges uniformly on all of U. The example $\{z^j\}$ on the unit disc D proves that assertion. The correct topology, for many purpose, on the space of holomorphic functions is uniform convergence on compact sets. See the book [LUR] for a more thorough treatment of this idea.

The proof just presented did not use the full force of the hypothesis that $|f(z)| \leq M$ for all $f \in \mathcal{F}$ and all $z \in U$. In fact, all we need is a bound of this sort for all z in each compact subset K, and the bound *may depend on* K. Thus we are led to the following definition.

Definition 3.7 Let $\mathcal{F}$ be a family of functions on an open set $U \subseteq \mathbb{C}$. We say that $\mathcal{F}$ is *bounded on compact sets* if, for each compact set $K \subseteq U$, there is a constant $M = M_K$ such that, for all $f \in \mathcal{F}$ and all $z \in K$:

$$|f(z)| \leq M.$$

The remarks in the preceding paragraph yield the proof of the following improved version of Theorem 3.5.

Prelude: Now we present a slighly more general version of Montel's result. This one is more natural, since the canonical topology on holomorphic functions is uniform convergence on compact sets.

THEOREM 3.8 (MONTEL'S THEOREM, SECOND VERSION) *Let $U \subseteq \mathbb{C}$ be an open set and let $\mathcal{F}$ be a family of holomorphic functions on U that is bounded on compact sets. Then, for every sequence $\{f_j\} \subseteq \mathcal{F}$, there is a subsequence $\{f_{j_k}\}$ that converges normally on U to a limit (necessarily holomorphic) function f_0.*

Example 3.9 We consider two instances of the application of Montel's theorem.

(A) Consider the family $\mathcal{F} = \{z^j\}_{j=1}^{\infty}$ of holomorphic functions. If we take U to be any subset of the unit disc, then $\mathcal{F}$ is bounded (by 1) so Montel's theorem (first version) guarantees that there is a subsequence that converges uniformly on compact subsets. Of course, in this case, it is plain by inspection that *any* subsequence will converge uniformly on compact sets to the identically zero function.

The family $\mathcal{F}$ fails to be bounded on compact sets for any U that contains points of modulus greater than 1. Thus neither version of Montel's theorem would apply on such a U. In fact, there is no convergent sequence in $\mathcal{F}$ for such a U.

(B) Let $\mathcal{F} = \{z/j\}_{j=1}^{\infty}$ on $\mathbb{C}$. Then there is no bound M such that $|z/j| \leq M$ for all j and all $z \in \mathbb{C}$. But for each fixed compact subset $K \subseteq \mathbb{C}$ there is a constant M_K such that $|z/j| < M_K$

for all j and all $z \in K$. (For instance, $M_K = \sup\{|z| : z \in K\}$ would do.) Therefore the second version of Montel's theorem applies. Indeed the sequence $\{z/j\}_{j=1}^{\infty}$ converges normally to 0 on $\mathbb{C}$.

Let us conclude this section by settling the derivative-maximizing holomorphic function issue that was raised earlier.

PROPOSITION 3.10 *Let $U \subseteq \mathbb{C}$ be any open set. Fix a point $P \in U$. Let $\mathcal{F}$ be a family of holomorphic functions from U into the unit disc D that take P to 0. Then there is a holomorphic function $f_0 : U \to D$ that is the normal limit of a sequence $\{f_j\}$, $f_j \in \mathcal{F}$, such that*

$$|f_0'(P)| \geq |f'(P)|$$

for all $f \in \mathcal{F}$.

Proof: We have already noted that the Cauchy estimates give a finite upper bound for $|f'(P)|$ for all $f \in \mathcal{F}$. Let

$$\lambda = \sup\{|f'(P)| : f \in \mathcal{F}\}.$$

By the definition of supremum, there is a sequence $\{f_j\} \subseteq \mathcal{F}$ such that $|f_j'(P)| \to \lambda$. But the sequence $\{f_j\}$ is bounded by 1 since all elements of $\mathcal{F}$ take values in the unit disc. Therefore Montel's theorem applies and there is a subsequence $\{f_{j_k}\}$ that converges uniformly on compact sets to a limit function f_0.

By the Cauchy estimates, the sequence $\{f_{j_k}'(P)\}$ converges to $f_0'(P)$. Therefore $|f_0'(P)| = \lambda$ as desired.

It remains to observe that, a priori, f_0 is known only to map U into the closed unit disc $\overline{D}(0,1)$. But the maximum modulus theorem implies that, if $f_0(U) \cap \{z : |z| = 1\} \neq \emptyset$, then f_0 is a constant (of modulus 1). Since $f_0(P) = 0$, the function f_0 certainly cannot be a constant of modulus 1. Thus $f_0(U) \subset D(0,1)$. □

3.2 Another Look at Normal Families

Capsule: There are many different ways to think about normal families. In the present section, we consider a differential geometric treatment. The key idea here is Marty's theorem.

Now we look at normal families from a differential geometric point of view.

Definition 3.11 The *spherical metric* is the natural metric on the unit sphere in $\mathbb{R}^3$ that is inherited from Euclidean space. When pushed down to the plane (by way of the stereographic projection), it takes the form

$$\frac{1}{1+|z|^2}\,dz\,d\overline{z}.$$

We denote the spherical metric by $|\ \ |_{\text{sph}}$. Now we have the following characterization, due to Marty, of normal families.

PROPOSITION 3.12 *Let $\mathcal{F} = \{f_\alpha\}$ be a family of holomorphic functions on a domain $U \subseteq \mathbb{C}^n$. Then $\mathcal{F}$ is a normal family if and only if, for each compact $K \subseteq U$, there is a constant $C = C_K > 0$ so that*

$$|f'(z)\cdot\xi|_{\text{sph}} \le C_K \cdot F_K^U(z;\xi) \tag{3.12.1}$$

for all $f \in \mathcal{F}$, all $z \in K$, and all $\xi \in \mathbb{C}$. Here F_K^U is the infinitesimal Kobayashi metric on U.

Proof: The "if" part of the assertion is obvious, for the inequality (3.12.1) gives a uniform bound on derivatives on any compact subset K of U. Thus we can apply the Ascoli-Arzela theorem as in the classical proof of Montel's theorem (see Section 3.1 above) and the result follows.

For the "only if" part, proceed as follows. Assume that the family $\mathcal{F}$ is normal, but that the inequality (3.12.1) fails. Then there are points $z^j \in K$ and Euclidean unit vectors $\xi^j \in \mathbb{C}^n$ and elements $f_j \in \mathcal{F}$ such that

$$|f_j'(z^j)\cdot\xi^j|_{\text{sph}} \ge j \cdot F_K^U(z^j,\xi^j). \tag{3.12.2}$$

Passing to a subsequence, and invoking the compactness of K and the compactness of the unit sphere in $\mathbb{C}^n$, we may suppose that $z^j \to z^0$, $\xi^j \to \xi^0$, and $f_j \to f_0$ uniformly on compacta.

Now, as $j \to \infty$, the left-hand side of (3.12.2) tends to $|f_0'(z^0)\cdot\xi^0|$. Also $F_K^U(z^j,\xi^j)$ tends to a number which is not less than F_K^U (z^0,ξ^0). However, if β is a large Euclidean ball containing U, then

$$0 < F_K^\beta(z^0,\xi^0) \le F_K^U(z^0,\xi^0),$$

which yields a contradiction. □

3.3 Normal Families in Their Natural Context

Capsule: It is natural to think of normal families of functions taking values in the Riemann sphere. These, of course, are meromorphic functions.

Thus we can begin to use the language of the spherical metric, which puts Marty's theorem in a natural context. As we look deeper, we begin to see connections with Picard's theorems.

Our usual definition of normal family makes more sense in a more general context. Instead of holomorphic, scalar-valued functions, let us consider *meromorphic functions.* A meromorphic function should be thought of as a holomorphic function taking values in the Riemann sphere $\widehat{\mathbb{C}}$. What does this mean?

The Riemann sphere $\widehat{\mathbb{C}}$ is a complex manifold. If

$$f : U \to \widehat{\mathbb{C}}$$

and $f(p) = \infty$, then we say that f is holomorphic near p if $1/f$ is holomorphic at p in the classical sense. [What is going on here, of course, is that we are applying a coordinate map to the image of f near the point ∞. See [LOS] for more on manifolds.] The important point here is that a sequence of functions

$$f_j : U \to \widehat{\mathbb{C}}$$

that converges uniformly on compact sets might in fact be compactly divergent. Consider the next example as an illustration.

Example 3.13 Let $f_j(z) = z^j$. Think of these as functions taking values in $\widehat{\mathbb{C}}$.

On the domain $U_1 = \{z \in \mathbb{C} : |z| < 1\}$, the sequence converges uniformly on compact sets to 0.

On the domain $U_2 = \{z \in \mathbb{C} : |z| > 1\}$, the sequence converges uniformly on compact sets to ∞ (if we think of the functions as taking values in $\widehat{\mathbb{C}}$). We may also say that, on U_2, the sequence is compactly divergent.

Now the elegant and all-inclusive formulation of Montel's theorem—in a temporarily heuristic format—is this:

Prelude: This version of Montel's theorem is a *pro forma* for some more precise versions that will be presented below.

THEOREM 3.14 (MONTEL) *Let $U \subseteq \mathbb{C}$ and let $\mathcal{F} = \{f_\alpha\}_{\alpha \in \mathcal{A}}$ be a family of meromorphic functions on U. If $\mathcal{F}$ satisfies a suitable growth or value-distribution condition then $\mathcal{F}$ is a normal family.*

This formulation of Montel's celebrated result is not very satisfying, just because it is too vague. The most naive condition to put on $\mathcal{F}$ is to mandate that $|f(z)| \leq M$ for some universal constant M and for all $f \in \mathcal{F}$, $z \in U$. But that would rule out compact divergence and would also rule out meromorphic functions. We need something more general. Even the more general condition, commonly found in texts (see [RUD2] or [GRK1]), that $|f(z)| \leq M_K$ for each compact set $K \subseteq U$ and some constant M_K depending on K would rule out compact divergence.

In fact, one of the most succinct formulations of Montel's theorem is this:

Prelude: This is a quite attractive version of Montel's result—clearly more general than the earlier versions. It also begins to suggest connections with Picard's theorems.

THEOREM 3.15 (MONTEL) *Let $\{\mathcal{F}\}$ be a family of meromorphic functions on a domain $U \subseteq \mathbb{C}$. If there are three values $\alpha, \beta, \gamma \in \widehat{\mathbb{C}}$ such that no $f \in \mathcal{F}$ takes any of the values α, β, γ, then $\mathcal{F}$ is a normal family.*

Note, in particular, that this new version of Montel includes as a special case the standard hypothesis that the family is uniformly bounded. But it includes a number of other interesting cases as well. It is not, however, a universal or all-inclusive result. For instance, it does not include the result about a family of functions that is uniformly bounded on compact sets.

There is just one unifying theorem in the subject, and that is Marty's theorem. We treat that result now.

As already indicated, it is most convenient to think of normal families in the context of *meromorphic* functions. This is just because we want to allow for convergence to infinity (formerly called "compact divergence"). Thus we will think of our functions as holomorphic maps from a domain $U \subseteq \mathbb{C}$ to the Riemann sphere $\widehat{\mathbb{C}}$. Since we want to look at things geometrically, we shall equip each of U and $\widehat{\mathbb{C}}$ with a metric. In fact, since uniform convergence on compact sets is a local property, we may as well take U to be the unit disc. We shall equip the disc with the Poincaré or Poincaré-Bergman metric. We shall equip the Riemann sphere with the spherical metric.

The latter metric may be unfamiliar, so let us take a few moments to discuss it. Imagine the stereographic projection p from the plane to the unit sphere—see Figure 3.2. For a point $(z, 0) = (x + iy, 0)$ in the plane, the intersection of the sphere with the line

$$t \mapsto (tx, ty, (1 - t))$$

is (along with the north pole $(0, 0, 1)$) the point

$$p(z) \equiv \left(\frac{2x}{|z|^2 + 1}, \frac{2y}{|z|^2 + 1}, \frac{|z|^2 - 1}{|z|^2 + 1} \right).$$

Now if $(z, 0) = (x + iy, 0)$ and $(w, 0) = (u + iv, 0)$ are points in the plane, then we measure their distance by taking the three-dimensional Euclidean distance between $p(z)$ and $p(w)$. The result is

$$\begin{aligned} s(z, w) &= \sqrt{\left(\frac{2x}{|z|^2 + 1} - \frac{2u}{|w|^2 + 1} \right)^2 + \left(\frac{2x}{|z|^2 + 1} - \frac{2v}{|w|^2 + 1} \right)^2} \\ &\qquad \overline{+ \left(\frac{|z|^2 - 1}{|z|^2 + 1} - \frac{|w|^2 - 1}{|w|^2 + 1} \right)^2} \\ &= \frac{2|z - w|}{\sqrt{1 + |z|^2}\sqrt{1 + |w|^2}}. \end{aligned}$$

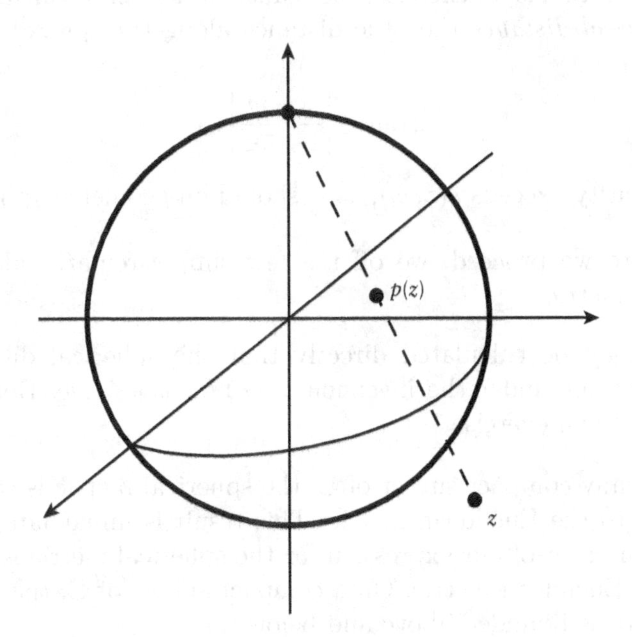

Figure 3.2: Stereographic projection.

This is the spherical metric. Some references call this the *chordal metric*.

It is also useful to have the infinitesimal form of the spherical metric. This can be calculated by pulling back the infinitesimal three-dimensional Euclidean metric under the stereographic projection, or by differentiating the metric that we just calculated. We take the latter approach. In point of fact,

$$\begin{aligned}\lim_{|h|\to 0}\frac{s(z+h,z)-s(z,z)}{|h|} &= \lim_{|h|\to 0}\frac{2|h|/[\sqrt{1+|z|^2}\sqrt{1+|z+h|^2}]-0}{|h|}\\ &= \lim_{|h|\to 0}\frac{2}{\sqrt{1+|z|^2}\sqrt{1+|z+h|^2}}\\ &= \frac{2}{1+|z|^2}.\end{aligned}$$

Thus if η is a tangent vector to $\widehat{\mathbb{C}}$ at the point w then

$$|\eta|_{\mathrm{sph},w} = \frac{2\|\eta\|}{1+\|w\|^2}.$$

Here $\|\cdot\|$ denotes the standard Euclidean length.

REMARK 3.16 As an exercise, the reader may wish to calculate that the *spherical distance* (i.e., the distance along the sphere) between $z, w \in \widehat{\mathbb{C}}$ is

$$2\tan^{-1}\left|\frac{z-w}{1+\overline{w}z}\right|.$$

This quantity exceeds $s(z,w)$, as it should on geometric grounds.

Before we proceed, we offer a few simple remarks about the spherical metric.

(1) It may be calculated directly that the spherical distance is invariant under the inversion $z \mapsto 1/z$, $w \mapsto 1/w$. Details are left as an exercise.

(2) On any compact subset of $\mathbb{C}$, the spherical metric is comparable to the Euclidean metric. This result is immediate, for the numerator of our expression for the spherical metric is already the Euclidean metric. On a compact subset of $\mathbb{C}$, the denominator is bounded above and below.

(3) On any compact subset of $(\mathbb{C}\setminus\{0\})\times(\mathbb{C}\setminus\{0\})$, we have the estimates

$$c\cdot\left|\frac{1}{z}-\frac{1}{w}\right| \le s(z,w) \le C\cdot\left|\frac{1}{z}-\frac{1}{w}\right|.$$

Here, c, C are positive constants. We invite the reader to verify this assertion.

(4) Our original definition of "uniform convergence on compact sets" may now be formulated in the language of the spherical metric. To wit, let $U \subseteq \mathbb{C}$ be a domain. A sequence $\{f_j\}_{j=1}^{\infty}$ of functions $f_j : U \to \widehat{\mathbb{C}}$ converges uniformly on compact sets to a limit function f (we sometimes say that the sequence $\{f_j\}$ converges *normally*) if $s(f_j(z), f(z))$ converges uniformly to 0 on compact subsets of U. Notice that this definition of normal convergence allows for the possibility of $f_j(z)$ converging to ∞ at certain values of z. In other words, the definition using the spherical metric amalgamates the ordinary notion of uniform convergence on compact sets with the newer notion of compact divergence.

Of course a holomorphic function $g : U \to \widehat{\mathbb{C}}$ is actually meromorphic. But we will sometimes find it convenient to refer to such functions as "holomorphic." No confusion should result if we keep the range of the function clearly in mind.

LEMMA 3.17 *Suppose that $f_j : U \to \widehat{\mathbb{C}}$ is a normally convergent sequence of holomorphic functions. Then the limit function f is a holomorphic function from U to $\widehat{\mathbb{C}}$ (in other words, the limit function is either meromorphic or identically ∞).*

Proof: At any point z where $f(z)$ is a finite complex number, we can be sure that there is a neighborhood U of z such that, for j sufficiently large, the f_j are uniformly bounded on U. But then f is necessarily a holomorphic function on U. At any point z where $f(z) = \infty$, we know that $1/f_j$ is uniformly bounded and converges uniformly to $1/f$ on some neighborhood U' of z. Now the same argument shows that $1/f$ is holomorphic in a neighborhood of z. In case $1/f \equiv 0$, then $f \equiv \infty$. Otherwise the zeros of $1/f$ are isolated; so f is meromorphic as claimed. □

We note that, in case the functions f_j in the lemma are actually holomorphic (not meromorphic), then the limit function f is either holomorphic or identically equal to ∞. The proof is essentially the same.

Now here is our philosophical approach to the matter of normal families. We begin with a sketch of the most classical Montel theorem. We have a family $\mathcal{F}$ of holomorphic functions on a domain U. The hypothesis is that $|f(z)| \leq M$ for all $f \in \mathcal{F}$ and all $z \in U$. Using the Cauchy estimates, this information implies that on a slightly smaller domain $\widetilde{U}$ we have $|f'(z)| \leq C \cdot M$ for all $z \in \widetilde{U}$. Now the mean value theorem tells us that, for $z \in \widetilde{U}$ and h small,

$$|f(z) - f(z+h)| \leq C \cdot M|h|$$

for all $f \in \mathcal{F}$. But this means that the family $\mathcal{F}$ is equicontinuous on $\widetilde{U}$. Of course $\mathcal{F}$ is equibounded by hypothesis. Now the Ascoli-Arzelà theorem applies to tell us that every sequence in $\mathcal{F}$ has a subsequence that converges uniformly on compact sets. That is the result.

The nub of this proof is the uniform bound on the derivative of any function $f \in \mathcal{F}$. This is the only step in the proof where "holomorphic" is used. Everything else in the proof is elementary topology and calculus. And this will be the inspiration for our formulation of Marty's result. We simply demand, for a family of meromorphic functions $f : D \to \widehat{\mathbb{C}}$, that the derivative be bounded from the Poincaré-Bergman metric on D (the unit disc) to the spherical metric on $\widehat{\mathbb{C}}$. Concretely, this condition is that

$$f^{\#}(z) \equiv \frac{2|f'(z)|}{1+|f(z)|^2} \leq C \cdot \frac{1}{1-|z|^2}.$$

In particular, this inequality means that $f^{\#}$ is bounded—independent of f—uniformly on compact subsets of D.

The expression $f^{\#}(z)$ is called the *spherical derivative* of f at the point z. Before we develop the theory any further, we make a few remarks about the spherical derivative.

1. If $f : U \to \widehat{\mathbb{C}}$ and $\gamma : [0,1] \to U$ is a continuously differentiable curve, then the spherical length of $f \circ \gamma$ is

$$\int_\gamma f^{\#}(z)|dz|.$$

2. The spherical derivative is invariant under inversion:

$$[1/f]^{\#}(z) = f^{\#}(z).$$

Now we come to the standard formulation of Marty's theorem as can be found in many standard complex function theory texts (see, for example, [GAM]). First we need a lemma.

LEMMA 3.18 *If f_j are meromorphic functions on the disc D and $f_j \to f$ uniformly on compact sets, then $f_j^\# \to f^\#$ uniformly on compact sets in D.*

Proof: This result follows as usual from the Cauchy estimates. □

Prelude: Marty's theorem is the classical, truly geometric version of Montel's theorem. It takes into account the spherical metric in a very natural way.

THEOREM 3.19 (MARTY) *Let $\mathcal{F}$ be a family of meromorphic functions on a domain $U \subseteq \mathbb{C}$. Then $\mathcal{F}$ is normal if and only if, for each compact set $K \subseteq U$ and $z \in K$, we have*

$$f^\#(z) \leq C_K,$$

where C_K is a constant that depends only on K.

Proof: Fix a point $p \in U$ and suppose that the spherical derivatives of elements of $\mathcal{F}$ are uniformly bounded on a neighborhood of p. Say that $f^\#(z) \leq C$ for $z \in D(p,r)$. If $q \in D(p,r)$ is any element and if γ is the straight line segment connecting p to q then, for any $f \in \mathcal{F}$, we may estimate the spherical distance from $f(p)$ to $f(q)$ by

$$s(f(p), f(q)) \leq \int_\gamma f^\#(z)\,|dz| \leq C \cdot |p-q|.$$

Thus we see that the family $\mathcal{F}$ is equicontinuous—from the Euclidean metric to the spherical metric. The Ascoli-Arzelà theorem now implies that $\mathcal{F}$ is normal.

Conversely, suppose that the spherical derivatives of the elements of $\mathcal{F}$ are *not* uniformly bounded on compact subsets of U. Then there are elements $f_j \in \mathcal{F}$ such that the maximum of $f_j^\#$ over some compact $K \subseteq U$ tends to $+\infty$. By the lemma, the sequence $\{f_j\}$ cannot have a normally convergent subsequence. □

Exercise for the Reader: Show that, under suitable hypotheses,

$$(g \circ f)^\#(z) = g^\#(f(z)) \cdot |f'(z)|.$$

Exercise for the Reader: Let γ be a compact curve in the complex plane that contains more than one point. Show that a family $\mathcal{F}$ of holomorphic functions on a domain U, taking values in $\mathbb{C} \setminus \gamma$, will be normal.

Exercise for the Reader: Suppose that $\mathcal{F}$ is a family of holomorphic functions $f : U \to \mathbb{C}$ and that

$$|f'(z)| \leq e^{|f(z)|}$$

for all $f \in \mathcal{F}$ and all $z \in U$. Then prove that $\mathcal{F}$ is a normal family.

3.4 Advanced Results on Normal Families

Capsule: One of the real pioneers in the modern studies of normal families has been L. Zalcman. He has a characterization of normal families that is closely related to Bloch's principle. We explore that idea here.

One of the unifying themes in the theory of normal families of functions is Lawrence Zalcman's result [ZAL] that follows. It is based on an old idea in function theory called the "Bloch principle." We shall see this technique used to good effect to establish a general version of Montel's theorem and thereby the Picard theorems. We shall also study Abraham Robinson's *Ansatz* about conditions that force a family of holomorphic functions to be normal.

PROPOSITION 3.20 *Assume that $\mathcal{F}$ is a family of meromorphic functions $f : U \to \widehat{\mathbb{C}}$ that is not normal. Then there exist:*

(a) *points $z_j \in U$ converging to a point $z \in U$;*

(b) *numbers $\rho_j > 0$ converging to 0;*

(c) *functions $f_j \in \mathcal{F}$ such that the dilated functions $g_j(\zeta) \equiv f_j(z_j + \rho_j\zeta)$ converge normally to a nonconstant meromorphic function g on $\mathbb{C}$ satisfying $g^{\#}(0) = 1$ and $g^{\#}(\zeta) \leq 1$ for all $\zeta \in \mathbb{C}$.*

Conversely, if these three conditions hold, then the family $\mathcal{F}$ is not normal.

Abraham Robinson's principle (which we shall discuss below) posits that there is a relationship between normal families and entire functions. Zalcman's lemma begins to make this connection palpable, for we see that a family that is not normal can be forced, after dilations and translations, to converge to an entire function with certain extremal properties.

Example 3.21 Zalcman's lemma is most useful in theoretical settings. But we shall present now a very simple example of how it works.

Let $f_j(z) = z^j$ and $\mathcal{F} = \{f_j\}_{j=1}^{\infty}$. Let $U = \{z \in \mathbb{C} : 1/2 < |z| < 2\}$. Then $\mathcal{F}$ is not normal.

Now select $z_j = 1$ for all j and $\rho_j = 1/j$. We plainly see that

$$\lim_{j\to\infty} f_j(z_j + \rho_j\zeta) = \lim_{j\to\infty}\left(1 + \frac{\zeta}{j}\right)^j = e^{\zeta} \equiv g(\zeta).$$

Furthermore,

$$g^{\#}(\zeta) = \frac{2|e^{\zeta}|}{1 + |e^{\zeta}|^2}.$$

Therefore $g^{\#}(0) = 1$ and (with a small calculation) $g^{\#}(\zeta) \leq 1$ for all ζ. We conclude that the family $\mathcal{F}$ is not normal.

Proof of Zalcman's Proposition: We may apply Marty's theorem to find a sequence $\{\alpha_j\}$ in some compact subset of U and functions $f_j \in \mathcal{F}$ such that $f_j^{\#}(\alpha_j) \to +\infty$. Translating coordinates, we may as well suppose that $\alpha_j \to 0 \in U$. Dilating coordinates, we may also assume that the closed unit disc $\overline{D}(0,1)$ lies in U.

Set

$$R_j \equiv \max_{|z|\leq 1}[f_j^{\#}(z)] \cdot (1 - |z|).$$

Since the sequence $\alpha_j \to 0$ and $f_j^{\#}(\alpha_j) \to +\infty$, we know immediately that $\lim_{j\to\infty} R_j = +\infty$. Suppose, for each j, that $[f_j^{\#}(z)] \cdot (1 - |z|)$ attains its maximum on the closed disc at the point β_j. Then

$$R_j = [f_j^{\#}(\beta_j)] \cdot (1 - |\beta_j|).$$

Since $f_j^{\#}(\beta_j) \geq R_j$, we may conclude that $f_j^{\#}(\beta_j) \to +\infty$.

Now we set

$$\rho_j = \frac{1}{f_j^{\#}(\beta_j)}.$$

Certainly $\rho_j > 0$ and $\rho_j \to 0$. Consider the disc parametrized by

$$\zeta \mapsto \beta_j + \rho_j\zeta, \quad |\zeta| < R_j. \tag{3.20.1}$$

This disc has center β_j and radius $1 - |\beta_j| = \rho_j R_j$. The disc defined in (3.20.1) lies in the unit disc and hence is a subset of U.

Define

$$g_j(\zeta) = f_j(\beta_j + \rho_j\zeta), \quad |\zeta| < R_j.$$

Since $R_j \to \infty$, the functions g_j are defined on larger and larger sets that exhaust $\mathbb{C}$. By the chain rule, we calculate that

$$g_j^\#(\zeta) = \rho_j \cdot f_j^\#(\beta_j + \rho_j\zeta), \quad |\zeta| < R_j.$$

Fix a positive number K. If j is so large that $R_j > K$, then $g_j(\zeta)$ is certainly defined on the disc $D(0, K)$. Since

$$[f_j^\#(\beta_j + \rho_j\zeta)] \cdot (1 - |\beta_j + \rho_j\zeta|) \leq R_j,$$

we may conclude that

$$\begin{aligned} g_j^\#(\zeta) &\leq \rho_j \cdot \frac{R_j}{1 - |\beta_j + \rho_j\zeta|} \\ &\leq \frac{\rho_j R_j}{1 - |\beta_j| - \rho_j K} \\ &= \frac{\rho_j R_j}{\rho_j R_j - \rho_j K} \\ &= \frac{1}{1 - K/R_j}, \quad |\zeta| < K. \end{aligned}$$

Marty's theorem now tells us that, on the disc $D(0, K)$, the functions $\{g_j\}$ form a normal family (we should restrict attention to j large so that all the functions are defined on $D(0, K)$). Passing to a subsequence, and re-indexing, we may thus assert that $\{g_j\}$ converges normally on $\mathbb{C}$ to a meromorphic function g. We know that, for each fixed K, the quantity $1/[1 - K/R_j]$ tends to 1. Hence the estimate on $g_j^\#$ shows that $g^\#(\zeta) \leq 1$ for all $\zeta \in \mathbb{C}$. Finally, because $g_j^\#(0) = \rho_j f_j^\#(\beta_j) = 1$, we know that $g^\#(0) = 1$. That concludes the proof of the first half of the result.

For the converse direction, suppose that $\mathcal{F}$ is a normal family. Say that $|z_j| < r$ for every j and some $r > 0$. By Marty's theorem, there is a constant $K > 0$ such that

$$\max_{|z| \leq [1+r]/2} f^\#(z) \leq K$$

for all $f \in \mathcal{F}$. Now suppose that

$$f_j(z_j + \rho_j\zeta) \to g(\zeta)$$

spherically uniformly on compact subsets of U. Fix a point $\zeta \in \mathbb{C}$. For j large, we know that $|z_j + \rho_j\zeta| \leq [1+r]/2$. Hence

$$\rho_j f^{\#}(z_j + \rho_j\zeta) \leq \rho_j K.$$

As a result, for all $\zeta \in \mathbb{C}$, we have

$$g^{\#}(\zeta) = \lim_{j\to\infty} \rho_j f_j^{\#}(z_j + \rho_j\zeta) = 0.$$

We conclude that g is constant (perhaps ∞). That concludes the proof of the converse direction. □

The following terminology will prove useful in our consideration of Picard's theorems.

Definition 3.22 Let f be meromorphic on a punctured disc $D'(p,r) \equiv D(p,r) \setminus \{p\}$ (we allow the possibility that f is actually holomorphic on $D(p,r)$). A value $\beta \in \widehat{\mathbb{C}}$ is called an *omitted value* of f if there is a number $\delta > 0$ such that $f(z) \neq \beta$ for all $0 < |z-p| < \delta$. An omitted value for f at the point $p = \infty$ is defined similarly.

Prelude: This is a generalization of the version of Montel that we presented in the previous theorem.

THEOREM 3.23 (MONTEL) *Let $\mathcal{F}$ be a family of meromorphic functions on a domain U. If there are extended complex numbers $\alpha, \beta, \gamma \in \widehat{\mathbb{C}}$ such that each $f \in \mathcal{F}$ omits each of α, β, γ, then the family $\mathcal{F}$ is normal.*

Proof: Normality is a local property, so we may as well suppose that $U = D(0,1)$. By postcomposing each element of $\mathcal{F}$ with a linear fractional transformation, we can assume that the omitted values are $0, 1, \infty$. Thus, in particular, the elements of $\mathcal{F}$ are actually holomorphic. Since these functions are nonvanishing on the simply connected domain D, they have roots of all orders. Define

$$\mathcal{F}_k = \{f^{1/2^k} \in \mathcal{F} : f \in \mathcal{F}\}.$$

The functions in $\mathcal{F}_k$ omit the values $0, 1$, and all 2^kth roots of unity. Evidently the family $\mathcal{F}_k$ is normal if and only if the family $\mathcal{F}$ is normal.

Seeking a contradiction, we now suppose that the family $\mathcal{F}$ is *not* normal. Thus, for each k, the family $\mathcal{F}_k$ is not normal. For each k, let g^k be the entire function for the family $\mathcal{F}_k$ that is produced by Zalcman's lemma. Thus $(g^k)^{\#}(0) = 1$ for each k and $(g^k)^{\#}(\zeta) \leq 1$ for all k and all ζ. Also, each g^k is the limit of scaled, translated functions in $\mathcal{F}_k$.

Since the family $\mathcal{F}_k$ omits the 2^kth roots of unity, then so do the translations, dilations, and restrictions of the elements of $\mathcal{F}_k$. By Hurwitz's theorem, so does any limit function. Hence each g^k omits the values that are the 2^kth roots of unity. By Marty's theorem, $\{g^k\}$ is a normal family (because the spherical derivatives are uniformly bounded). Let g^0 be the normal limit of some subsequence. By Hurwitz's theorem, g^0 will omit *all* 2^kth roots of unity for *all* k (it also omits 0 and ∞). Because g^0 is an open mapping, it must be therefore that g^0 omits every value on the unit circle.

Thus the image of g^0, which is certainly connected, lies either in $D(0,1)$ or in $\mathbb{C} \setminus \overline{D(0,1)}$. In the first instance—since g^0 is entire—we see that g^0 is constant. In the second instance, the function $1/g^0$ (remember that g^0 does not vanish!) is constant. This is a contradiction because certainly $(g^0)^{\#}(0) = 1$. □

Recall that the Casorati-Weierstrass theorem tells us that the distribution of values of an analytic function near an essential singularity is dense in the complex numbers. The theorems of Picard strengthen this result in a decisive fashion.

Prelude: As we predicted, this circle of ideas is related to Picard's theorems. The idea of omitting three values is geometrically significant. In fact, there is a proof of Picard (due to B. Davis) that uses Brownian motion and makes the significance of "three" quite clear.

THEOREM 3.24 (PICARD'S GREAT THEOREM) *Suppose that f is holomorphic on a punctured disc $D'(p,r) \equiv D(p,r) \setminus \{p\}$. If f has an essential singularity at p, then f cannot omit two values at p.*

Proof: We may as well suppose that $p = 0$ and that the function omits the two values 0 and 1. We shall establish then that f cannot have an essential singularity at $p = 0$.

Let $\delta_1 > \delta_2 > \cdots \to 0$. Set

$$f_j(z) = f(\delta_j z), \quad 0 < |z| < r.$$

Then $\{f_j\}$ omits three values (the two hypothesized plus ∞). Hence, by Montel, $\{f_j\}$ is a normal family. Let f^0 be a subsequential limit on $D'(p,r)$.

If f^0 is not identically ∞, then f^0 is holomorphic on $D'(p, r)$. Let $0 < s < r$. There is a positive number M such that $|f^0(z)| < M$ when $|z| = s$. For j large, we may then conclude that $|f_j(z)| < M$ when $|z| = s$. In conclusion, $|f(z)| < M$ for $|z| = \delta_j s$. By the maximum principle, $|f(z)| < M$ for $\delta_j s \leq |z| \leq s$, $j = 1, 2, \ldots$. Taking the union over j, we find that $|f(z)| < M$ on the punctured disc $D'(p, s)$. Now the Riemann removable singularities theorem implies that f continues analytically across the puncture at p. That is a contradiction.

If instead f^0 *is* identically ∞, then we simply apply the preceding argument to the function $1/f^0$. The conclusion is that $1/f$ continues analytically across p with value 0 at p. Thus f has a pole at p. That is a contradiction. □

Prelude: Picard's little theorem is a weaker result than the great theorem. It is still quite interesting and useful.

THEOREM 3.25 (PICARD'S LITTLE THEOREM) *A nonconstant entire function f cannot omit two values.*

Proof: If f is a polynomial, then f takes *all* values by the fundamental theorem of algebra.

If f is transcendental, then f has an essential singularity at ∞. Now apply the result of the preceding theorem (exercise for the reader). □

It is easy to find examples that illustrate the principle of Picard's theorems. First, a polynomial is entire and omits no values. The function $f(z) = e^z$ is entire and omits just one value. The function $f(z) = e^{1/z}$ has an essential singularity at the origin and omits the value 0 near the origin. The function ze^z is entire and omits no values. The function $\cos z$ is entire and omits no values.

3.5 Robinson's Heuristic Principle

Capsule: Abraham Robinson was a remarkably diverse mathematician. He is the person who created the nonstandard real numbers—thus finally putting the idea of infinitesimal on a rigorous footing.

In an important lecture, he formulated an heuristic idea of why a family of holomorphic functions should be normal. This idea is closely related to the results of Zalcman that we explored in the last section.

In his retiring presidential address to the Association for Symbolic Logic [ROB], Abraham Robinson proposed an intuitive condition for normality of a family of holomorphic or meromorphic functions. It says that any "property" that would cause an entire function to be constant would also cause a family of holomorphic or meromorphic functions to be normal. If we ignore, for the moment, the fact that we do not have a rigorous definition of "property," then we can certainly appreciate Robinson's idea. For example, "boundedness" will cause an entire function to be constant. It will also, by Montel's theorem, cause a family of holomorphic functions to be normal. Omission of two values will cause an entire function to be constant, and it will also cause a family of holomorphic functions to be normal.

In the present section, we shall consider Robinson's principle. We shall follow the presentation in [ZAL], which gives the principle a rigorous formulation and a proof. In what follows, we will consider (instead of functions in the ordinary sense) *function elements.* A function element is an ordered pair (f, U), where $U \subseteq \mathbb{C}$ is an open set and f a holomorphic or meromorphic function on U. A *property* P is a collection of function elements.

Definition 3.26 Let P be a property that satisfies the following conditions:

(a) If $(f, U) \in P$ and $\widetilde{U} \subseteq U$, then $(f, \widetilde{U}) \in P$.

(b) If $(f, U) \in P$ and $\phi(\zeta) = a\zeta + b$ is an affine holomorphic mapping, then $(f \circ \phi, \phi^{-1}(U)) \in P$.

(c) Let $(f_j, U_j) \in P$, where $U_1 \subseteq U_2 \subseteq \cdots$, and set $U = \cup_j U_j$. If $f_j \to f$ uniformly on compact sets in the spherical metric, then $(f, U) \in P$.

We call such a property *critical.*

Prelude: This theorem is a precise enunciation of A. Robinson's heuristic principle for normal families. It is particularly interesting because of the definition of "property."

THEOREM 3.27 *Let P be a critical property. Assume that $(f, \mathbb{C}) \in P$ if and only if f is a constant function. Then, for any domain U, the family of functions f satisfying $(f, U) \in P$ is normal.*

This theorem is a rigorous enunciation of Abraham Robinson's idea. According to Zalcman [ZAL], it is the outgrowth of ideas of

Pommerenke. We shall now prove the result and then offer some examples and applications. The reference [ALK] offers some generalizations of Zalcman's ideas.

Proof of Theorem 3.27: Let $\mathcal{F}$ be the family of all functions on a domain U that has property P. If $\mathcal{F}$ is not normal, then Marty's condition shows that it is not normal on some subdisc.

After a change of coordinates, we may as well suppose that that disc is the unit disc $D(0,1) = D$. Use the notation from the statement and proof of Zalcman's proposition. Define $R_j = (r - |z_j|)/\rho_j$. Since $R_j \to \infty$, we may suppose (passing to a subsequence if necessary) that the R_js form an increasing sequence. Let $g_j(\zeta) = f_j(z_j + \rho_j\zeta)$ and $D_j = D(0, R_j) = \{\zeta \in \mathbb{C} : |\zeta| < R_j\}$. The functions (g_j, D_j) satisfy Property P by conditions **(a)** and **(b)**. Now condition **(c)** implies that $(g, \mathbb{C})$ also satisfies Property P. Since P has no nonconstant functions that are defined on all of $\mathbb{C}$, we now have a contradiction. It follows that $\mathcal{F}$ must be a normal family. □

It is important now that we illustrate Robinson's heuristic principle with a few applications. We begin with a new proof of Montel's theorem.

Prelude: Now we can use A. Robinson's ideas to give a new proof of Montel's theorem.

THEOREM 3.28 (MONTEL) *Let $\mathcal{F}$ be a family of meromorphic functions on the domain U. If $\mathcal{F}$ omits the three values α, β, γ, then $\mathcal{F}$ is a normal family.*

Proof: We apply Robinson's principle. Take for Property P the condition "either f is constant or it omits the values α, β, γ on U." We see immediately that properties **(a)** and **(b)** of "critical property" hold. Also, condition **(c)** is a consequence of Hurwitz's theorem. Of course we know that any meromorphic function on all of $\mathbb{C}$ that satisfies Property P must be constant—that is Picard's Great Theorem. The proof is complete. □

We confess that our treatment here is not entirely satisfactory. After all, we used Montel's theorem earlier in the chapter to *prove* Picard's Great Theorem. Now we are using Picard to prove Montel. The concerned reader may consult [GRK1], for example, where Picard is proved using the elliptic modular function, and thus in a fashion that is independent of the present discussion. The book [KRA1] presents yet another point of view.

Zalcman [ZAL] presents a rather exotic version of Montel's theorem, which is not very well known. We recount it here.

Prelude: This very interesting version of Montel's theorem generalizes the "three omitted values" version in a dramatic way. The proof uses the spherical metric decisively.

THEOREM 3.29 *Let $\mathcal{F}$ be a family of meromorphic functions on the domain U. Assume that each function $f \in \mathcal{F}$ omits three distinct values—but the choice of the three values may depend on f. Call the omitted values $a = a_f$, $b = b_f$, and $c = c_f$. Suppose that s is the spherical distance and there is a constant $\lambda > 0$ (independent of $f \in \mathcal{F}$) such that we always have*

$$s(a,b) \cdot s(b,c) \cdot s(c,a) \geq \lambda > 0.$$

Then $\mathcal{F}$ is a normal family.

Proof: Let P be the property "f omits three values a, b, c such that $s(a,b) \cdot s(b,c) \cdot s(c,a) \geq \lambda$." By Picard's theorem, no nonconstant meromorphic function can enjoy Property P. Certainly properties **(a)** and **(b)** of "critical property" are satisfied. The proof will be complete if we can show that Property P is preserved under uniform convergence in the spherical metric.

So let us say that $f_j \to f$ spherically uniformly on compact subsets of U and that each f_j omits the values a_j, b_j, c_j with $s(a_j, b_j) \cdot s(b_j, c_j) \cdot s(c_j, a_j) \geq \lambda > 0$. We may suppose that f is nonconstant, for otherwise it trivially satisfies Property P. Now the sphere is compact, so there exist points $a, b, c \in \widehat{\mathbb{C}}$ and a subsequence—still indexed by j—such that $s(a_j, a) \to 0$, $s(b_j, b) \to 0$, and $s(c_j, c) \to 0$. Continuity implies that $s(a,b) \cdot s(b,c) \cdot s(c,a) \geq \lambda$. We shall show that f omits the values a, b, c.

If $f(w) = a$ and $a \neq \infty$, then select a number $r > 0$ such that $K = \{z \in \mathbb{C} : |z - w| \leq r\} \subseteq U$ and f is holomorphic on K. Of course f is bounded on K, so $f_j(z) - a_j \to f(z) - a$ on K. The function $f(z) - a$ is nonconstant and vanishes at $z = w$. Hurwitz's theorem then implies that $f_j(z) - a_j$ must vanish on K for large j. That is a contradiction.

If instead $a = \infty$, then consider the functions $1/f_j$, $1/f$. Arguing as before, and using the invariance of the spherical metric under inversion, we see that the proof is complete. □

Chapter 4

Automorphism Groups

Prologue: In 1906, Henri Poincaré proved that the unit ball B in $\mathbb{C}^2$ and the unit bidisc D^2 in $\mathbb{C}^2$ are *not* biholomorphically equivalent. He proved this profound result by calculating the automorphism group (group of biholomorphic self maps) of each domain and showing that those automorphism groups cannot be isomorphic.

This was an early and profound use of the theory of automorphism groups. But Poincaré's theorem has had a profound influence on complex analysis. It tells us that there is no Riemann Mapping theorem in several complex variables. As a result, we seek means to compare and to contrast domains in $\mathbb{C}^n$. One of the key tools in doing so is the theory of automorphism groups.

What is interesting about the automorphism group of a bounded domain in $\mathbb{C}^n$ is that it is a (real) Lie group. Thus the study of automorphism groups involves both complex analysis and real analysis (and algebra). There are even papers in the subject that use formal logic. This confluence of many different fields produces a rich theory that is full of important ideas and many surprises.

In modern times, the theory of automorphism groups has taken on a life of its own. It is used to understand the geometry of complex domains, but it is also studied in its own right. Automorphism groups have a meaning-ful and profound connection with Bergman geometry—a topic that we treat later in this book. They also can be profitably studied using the Carathéodory and Kobayashi metrics.

In this chapter, we give the reader a basic introduction to the theory and practice of automorphism groups. We are able to prove a few substantial results, and to point in some further directions.

Felix Klein's *Erlangen Program* lays out a blueprint for understanding a geometry by way of the group of mappings that preserve that geometry. This vision has become quite prevalent and powerful in modern approaches to the subject. Certainly Alexandre Grothendieck and Saunders Mac Lane carried this idea to new heights in their modern formulations of algebraic geometry and algebraic topology.

For complex function theory, Klein's idea may be implemented by means of the study of conformal mappings. While there is certainly value in studying individual mappings, there is even more to be had from studying *groups* of conformal mappings. Thus we are led to study the group of conformal self-maps (the *automorphism group*) of a planar domain. This idea interacts elegantly and effectively with the notion of invariant metric (see Chapter 2).

A very basic acquaintance with the group concept and with conformal mappings is all that is required to appreciate the ideas in the present chapter. The study of automorphism groups exhibits a fruitful interaction of algebra, analysis, and geometry. It points to many new directions in the subject.

4.1 Introductory Concepts

Capsule: In this section we introduce some foundational ideas about the automorphism group. In particular, we treat some classical ideas of H. Cartan.

We begin to see how analysis of the automorphism group is accomplished, and some of the techniques that come into play.

As mentioned earlier, Felix Klein taught us that a natural way to understand a geometric object is by way of the groups of transformations that act on that object. In complex analysis, the natural transformations (or, in more abstract language, the "morphisms") are the conformal (one-to-one, onto, holomorphic) mappings. If $\phi : U_1 \to U_2$

is such a mapping, then ϕ is a device for transplanting the complex analysis of U_1 to the complex analysis of U_2 (and vice versa).

If U is a fixed domain, then of particular interest are the conformal *self-maps* of U. The collection of such mappings forms a group under composition of mappings:

- The composition of two self-maps is another,
- Each self-map of U has an inverse that is also a self-map,
- The identity map is the group identity,
- The binary operation of composition of mappings is associative.

We call this group the *automorphism group* of U, denote it by $\text{Aut}\,(U)$, and call its elements *automorphisms.*

The following result of H. Cartan will be of considerable utility in our study of automorphisms:

PROPOSITION 4.1 *Let $U \subseteq \mathbb{C}$ be a bounded domain. Let $P \in U$, and suppose that $\phi : U \to U$ satisfies $\phi(P) = P$. If $\phi'(P) = 1$, then ϕ is the identity.*

Proof: We may assume that $P = 0$. Expanding ϕ in a power series about $P = 0$, we have

$$\phi(z) = z + P_k(z) + O(|z|^{k+1}),$$

where P_k is the first nonvanishing monomial (of degree k) of order exceeding 1 in the Taylor expansion. Notice that

$$P_k(z) = \left(\frac{1}{k!}\frac{\partial^k}{\partial z^k}\phi(0)\right) z^k.$$

Defining $\phi^j(z) = \phi \circ \cdots \circ \phi$ (j times) we have

$$\begin{aligned}
\phi^2(z) &= z + 2P_k(z) + O(|z|^{k+1}),\\
\phi^3(z) &= z + 3P_k(z) + O(|z|^{k+1}),\\
&\vdots\\
\phi^j(z) &= z + jP_k(z) + O(|z|^{k+1}).
\end{aligned}$$

Choose discs $D(0, a) \subseteq U \subseteq D(0, b)$. Then for $0 \leq j \in \mathbb{Z}$ we know that $D(0, a) \subseteq \text{dom}\,\phi^j \subseteq D(0, b)$. Therefore the Cauchy estimates imply that, for our index k, we have

$$j\left|\frac{\partial^k}{\partial z^k}\phi(0)\right| = \left|\frac{\partial^k}{\partial z^k}\phi^j(0)\right| \leq \frac{b \cdot k!}{a^k}.$$

Letting $j \to \infty$ yields that $\phi^{(k)}(0) = 0$.

We conclude that $P_k = 0$; this contradicts the choice of P_k unless $\phi(z) \equiv z$. □

There are other ways to look at this result, some of them quite illuminating. Imagine that U is equipped with the Bergman metric (see Chapter 9 below). Because $\phi'(P) = 1$, we see that ϕ is invertible in a neighborhood of P. If γ is a geodesic emanating from P, then $\phi \circ \gamma$ will also be a geodesic emanating from P. In fact, since $\phi'(P) = 1$, the curve $\phi \circ \gamma$ will be a geodesic emanating from the same point and pointing in the same direction. By uniqueness for first-order differential equations (Picard's theorem), $\phi \circ \gamma$ will coincide with γ—at least at points near P. Since this statement is true for *any* geodesic emanating from P, we may conclude that ϕ is the identity in a neighborhood of P. By analytic continuation, we see that $\phi(z) \equiv z$.

Now suppose that U is a fixed domain and that $P \in U$ is a given point. Let ϕ_1, ϕ_2 be automorphisms of U such that $\phi_1(P) = \phi_2(P)$ and $\phi_1'(P) = \phi_2'(P)$. Then $\psi \equiv \phi_1 \circ \phi_2^{-1}$ has the property that $\psi : U \to U$, $\psi(P) = P$, and $\psi'(P) = 1$. By Cartan's result, we conclude that $\psi(z) \equiv z$ and hence $\phi_1(z) \equiv \phi_2(z)$. In conclusion, the mapping

$$\mathrm{Aut}\,(U) \ni \phi \mapsto (\phi(P), \phi'(P))$$

is one-to-one. In effect, this mapping shows that the automorphism group may be identified with a subset of some Euclidean space (the target is $\mathbb{C} \times \mathbb{C}$, which is $\mathbb{R}^4$). In point of fact (we shall not provide the details here, but see [KOB]), Hilbert's fifth problem tells us that the group $\mathrm{Aut}\,(U)$ is a *Lie group*, which simply means that $\mathrm{Aut}\,(U)$ is a group that is also a manifold, and the group operations are continuous (indeed, real analytic) in the topology of the manifold.[1]

One consequence of the discussion in the last paragraph is that, for a given U, the group $\mathrm{Aut}\,(U)$ has a *dimension*. It is natural to wonder what the dimension of $\mathrm{Aut}\,(U)$ tells us about U itself.

4.2 Noncompact Automorphism Groups

Capsule: It is natural to consider domains that have lots of automorphisms. A domain U is said to have transitive automorphism group if, whenever P, $Q \in U$, then there is an automorphism φ so

[1] These issues are related to Hilbert's fifth problem, which asked whether any locally Euclidean group is a Lie group. See [BRO] and [TAU].

that $\varphi(P) = Q$. It turns out, however, that domains with transitive automorphism group are rather rare. In fact the only example is the disc.

As a result, we instead consider domains with noncompact automorphism group. This idea is geometrically natural, and is a logical generalization of transitive automorphism group. A domain with noncompact automorphism group has plenty of automorphisms, and always has an interior point with orbit that runs out to the boundary. This gives us a way to relate the boundary geometry of the domain to the global geometry of the domain.

Before we can engage in a detailed analysis of the dimensions of automorphism groups, we must amass some other ideas and techniques.

We have already noted that the automorphism group is a topological group. In fact the topology we use, as is standard in complex analysis, is the topology of *uniform convergence on compact sets* (in point-set topology this is known as the *compact-open topology*). As we shall see in the developing ideas, it is a matter of some interest to see when the automorphism group is compact and when it is not. First we consider some examples.

Example 4.2 Let $D = \{z \in \mathbb{C} : |z| < 1\}$ and $A = \{z \in \mathbb{C} : 1/2 < |z| < 2\}$. Then D is the unit disc in the complex plane and A is an annulus. Let us use the idea of the automorphism group to see that D and A cannot be conformally equivalent.[2]

Now we know from any complex analysis text (see, for example, [GRK1]), that the collection of conformal self-maps of the disc D consists of all maps of the form

$$\zeta \mapsto e^{i\tau} \cdot \frac{\zeta - a}{1 - \overline{a}\zeta},$$

where τ is a fixed real number and a is a complex number of modulus less than 1. On the other hand, the collection of conformal maps of the annulus A consists of all rotations $\zeta \mapsto e^{i\tau}\zeta$ together with the inversion $\lambda : \zeta \mapsto 1/\zeta$ (and compositions of these two types). Again see [GRK1].

[2]It should be stressed that it is a priori clear that these two domains cannot be conformally equivalent because they are not even topologically equivalent. After all, D is simply connected while A is not. But our point is to see how automorphism groups can be used to obtain useful results.

Consider the elements $\alpha_j \in \text{Aut}(D)$ given by

$$\alpha_j(\zeta) = \frac{\zeta - (1 - 1/j)}{1 - (1 - 1/j)\zeta}.$$

Then we see that

$$\lim_{j \to \infty} \alpha_j(\zeta) \equiv -1,$$

uniformly on compact sets. We conclude that the topological group Aut (D) is *not* compact; for we have produced a sequence of elements of Aut (D) that has no subsequence *converging to an element of Aut* (D).

By contrast, our explicit description of the automorphism group of A shows that Aut (A) is just two copies of the circle. Put more analytically, if β_j are elements of Aut (A), then either **(i)** infinitely many of them have a factor of the inversion λ or else **(ii)** infinitely many of them do not. Suppose the former. Then each β_j may be written $\beta_j = \lambda \circ e^{i\rho_j}$, where $0 \le \rho_j \le 2\pi$. Then it is easy to extract a subsequence that converges uniformly. As a result, Aut (A) is compact. [A similar argument applies in the second eventuality.]

Now, if there were a conformal mapping $\psi : D \to A$, then the induced mapping

$$\begin{aligned} \psi_* : \text{Aut}(D) &\to \text{Aut}(A), \\ \phi &\mapsto \psi \circ \phi \circ \psi^{-1}, \end{aligned}$$

would be a topological group isomorphism. But these two groups *cannot* be isomorphic as topological groups just because one is compact and the other is not![3]

We conclude that D and A are not conformally equivalent.

Thus we see that the automorphism group can serve as a device for differentiating conformal equivalence and/or inequivalence of domains. In the context of the plane, this device may seem somewhat artificial, just because so many other powerful tools (the Riemann Mapping theorem, the uniformization theorem, tools from Riemann surface theory, etc.) are at our disposal.

Compactness/noncompactness is only one of many possible dialectics that we could use to see that the annulus and the disc cannot be conformally equivalent. We could also look at the *dimension*

[3]It is also worth noting that one automorphism group (for the annulus) is abelian while the other (for the disc) is not. Details are left for an exercise.

of the automorphism groups. As already noted, the automorphism group of the annulus is two circles, and has dimension one. By contrast, the automorphism group of the disc is all maps of the form

$$\zeta \mapsto e^{i\tau} \cdot \frac{\zeta - a}{1 - \overline{a}\zeta}.$$

Plainly, there are three free parameters (one for the real parameter τ and two for the complex parameter a) and the automorphism group of the disc therefore has dimension three. As a result, the annulus and the disc cannot be conformally equivalent.

We may apply this philosophy in other interesting circumstances. Part of the content of the uniformization theorem of Koebe (see Section 1.8) is that there are only three (conformally distinct) simply connected Riemann surfaces: the disc D, the plane $\mathbb{C}$, and the Riemann sphere $\widehat{\mathbb{C}}$. Let us discuss for a moment the automorphism groups of these objects.

As already noted, Aut (D) has dimension three. The conformal self-mappings of the plane are those of the form

$$\zeta \mapsto a\zeta + b$$

for arbitrary complex constants $a \neq 0$ and b. Thus Aut $(\mathbb{C})$ has dimension four. The conformal self-mappings of the sphere are the *linear fractional transformations*

$$\zeta \mapsto \frac{a\zeta + b}{c\zeta + d}.$$

To eliminate redundancies, we write these (by dividing out by c or d, whichever is nonzero) as

$$\zeta \mapsto \frac{\alpha\zeta + \beta}{\gamma\zeta + 1} \quad \text{or} \quad \zeta \mapsto \frac{\alpha\zeta + \beta}{\zeta + \delta}.$$

Thus, plainly, the dimension[4] of Aut $(\widehat{\mathbb{C}})$ is 6. In any event, the calculations in the last paragraph explain why the disc, the plane, and the sphere must all be conformally distinct. (The sphere must be distinct from the other two anyway because it is topologically inequivalent.)

[4]It is worth noting that the conformal self-maps of the plane are just those conformal self-mappings of the sphere that map ∞ to ∞. When we pass from mappings of the plane to mappings of the sphere, we have the extra latitude of moving the point at ∞ to another point on the sphere—and that gives two more degrees of freedom. That is why $\dim(\text{Aut}\,(\widehat{\mathbb{C}})) = 6$ and $\dim(\text{Aut}\,(\mathbb{C})) = 4$.

Note that a more classical proof of the conformal inequivalence of the disc and the plane follows from Liouville's theorem.

Given what we have seen so far, it is natural to wonder which planar domains have compact automorphism group. It is a classical fact (see [HEI1], [HEI2], or [FAK]) that any domain with at least two, but finitely many, holes has in fact a *finite* automorphism group. (We shall provide some heuristic justification for this assertion below.) Thus the automorphism group will certainly be compact. It is natural, then, to conjecture that if a planar domain is *not* simply connected, then it will have compact automorphism group (and, conversely, if the planar domain *is* simply connected, then it will have noncompact automorphism group).

Part of this statement is obvious: If U is a simply connected planar domain, not the entire plane, then the Riemann Mapping theorem tells us that U is conformally equivalent to the disc D. But then the argument given in Example 4.2 shows that Aut (U) will be topologically isomorphic to Aut (D). It follows, then, that Aut (U) is noncompact.

The other part of the statement is *false*, as the following example shows.

Example 4.3 Let $D = \{\zeta \in \mathbb{C} : |\zeta| < 1\}$ as before. Let $\overline{d} = \{\zeta \in \mathbb{C} : |\zeta| \leq 1/10\}$. Define

$$\phi(\zeta) = \frac{\zeta - 1/2}{1 - (1/2)\zeta}.$$

For $1 \leq j \in \mathbb{Z}$, we let

$$\phi^j(\zeta) \equiv \underbrace{\phi \circ \phi \circ \cdots \circ \phi(\zeta)}_{j \text{ times}}.$$

If $j < 0$, $j \in \mathbb{Z}$, we let

$$\phi^j(\zeta) \equiv [\phi^{-1}]^{|j|}(\zeta).$$

Finally, we set

$$\phi^0(\zeta) = \zeta.$$

With this language, we define

$$U = D \setminus \bigcup_{j=-\infty}^{\infty} \phi^j(\overline{d}).$$

Figure 4.1 exhibits this domain. It is obviously infinitely connected. What is its automorphism group?

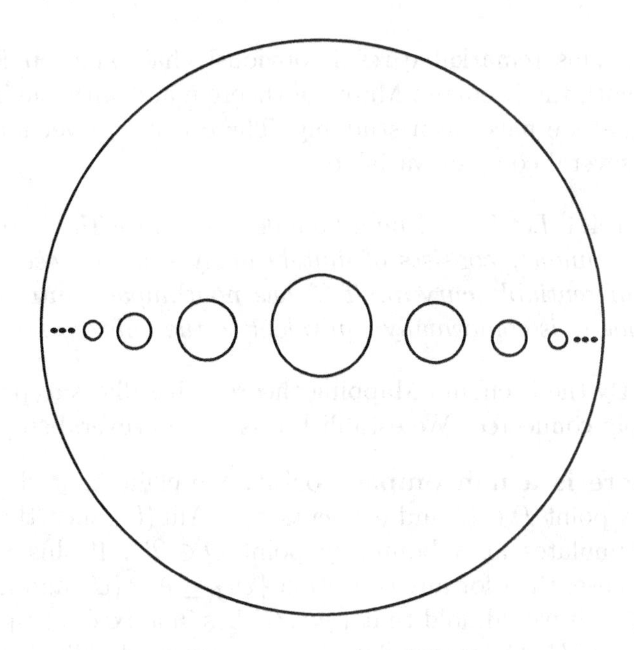

Figure 4.1: The domain U.

Certainly any ϕ^k is an automorphism of U, just because $\phi^k \circ \phi^j = \phi^{j+k}$. Also the mapping $\zeta \mapsto -\zeta$ is an automorphism of U. It is not difficult to show, though we shall not provide the details, that these are all the automorphisms of U.

Further note that, by dint of tedious hand calculation, we may actually calculate ϕ^j. In fact, for all j,

$$\phi^j(\zeta) = \frac{\zeta - \frac{3^j-1}{3^j+1}}{1 - \frac{3^j-1}{3^j+1}\zeta}.$$

We see, therefore, that $\phi^j \to -1$, uniformly on compact sets, as the sequence $j \to +\infty$ (and likewise $\phi^j \to +1$, uniformly on compact sets, as $j \to -\infty$). In conclusion, Aut (U) is noncompact.

The upshot of this last example is that the possession of holes is not the sole determinant of when a planar domain has compact automorphism group. In fact part of what makes this subject fascinating is that the positive results require a combination of algebra, complex function theory, and differential geometry. The next theorem answers the quandary into which we have fallen, and also illustrates this cross-fertilization of techniques.

Prelude: This remarkable result obviously has philosophical connections with the Riemann Mapping theorem and with the invariant metrics that we have been studying. There is also a version of this result in several complex variables.

THEOREM 4.4 *Let $U \subseteq \mathbb{C}$ be a bounded domain with C^1 boundary (i.e., the boundary consists of finitely many simple, closed, continuously differentiable curves). If U has noncompact automorphism group, then U is conformally equivalent to the unit disc.*

Proof: By the Riemann Mapping theorem, it suffices to prove that U is simply connected. We establish this fact in several steps.

(1) There is a noncompact orbit. We claim that there must be a point $P \in U$ and elements $\phi_j \in \mathrm{Aut}\,(U)$ such that $\phi_j(P)$ accumulates at a boundary point $Q \in \partial U$. If this were not the case, then for any collection $\{\psi_j\} \subseteq \mathrm{Aut}\,(U)$ and any fixed $P \in U$ it would hold that $\{\psi_j(P)\}$ lies in a fixed, compact subset $L \subseteq U$. Of course $\mathrm{Aut}\,(U)$ is a normal family (since U is bounded); hence there is a normally convergent subsequence ψ_{j_k} that converges to some holomorphic limit function ψ_0. Now the argument principle shows easily that this limit function ψ_0 is univalent. The fact that all the $\psi_{j_k}(P)$ are trapped in L shows that the limit function maps U into U. Since similar reasoning applies to the inverse mappings, we may conclude that the limit mapping ψ_0 is an automorphism. So we see that $\mathrm{Aut}\,(U)$ is compact. Since we assumed that this is not the case, we may therefore conclude that the asserted points $P \in U$ and $Q \in \partial U$ and automorphisms ψ_j must exist.

(2) There is a peaking function at the point $Q \in \partial U$. A "peaking function" is a function $\mu : \overline{U} \to \overline{D}$ such that

- μ is continuous on $\overline{U}$.
- μ is holomorphic on U.
- $\mu(Q) = 1$.
- $|\mu(\zeta)| < 1$ for all $\zeta \in \overline{U} \setminus \{Q\}$.

Such a peaking function may be constructed by mapping U univalently, with a mapping ξ, into the unit disc so that the holes in U go to interior domains in the disc and the outer boundary of U goes to the boundary of the disc. Carathéodory's theorem

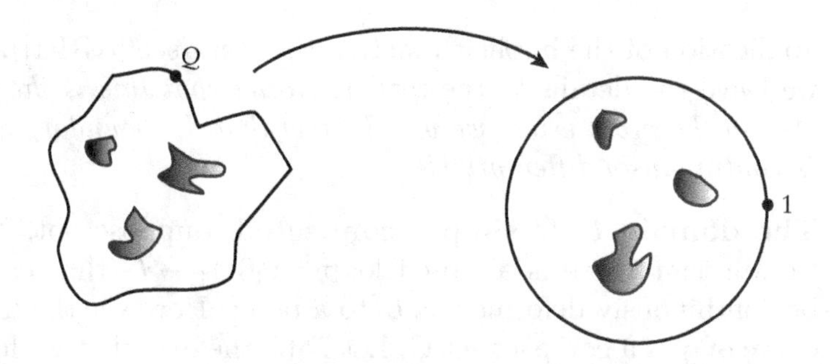

Figure 4.2: Continuous and univalent extension of the mapping.

tells us that the mapping extends continuously and univalently to the boundary. See Figure 4.2. Suppose that the boundary point $Q \in \partial U$ gets mapped to $1 \in \partial D$. Now the function

$$f(\zeta) = \frac{\zeta + 1}{2}$$

is a peaking function for the point 1 in ∂D. Thus $\mu \equiv f \circ \xi$ is the peaking function that we seek for U at Q.

(3) **If $K \subseteq U$ is any compact subset and if ψ_j and P, Q are as in assertion (1), then there is a subsequence ψ_{j_k} such that $\psi_{j_k}(z) \to Q$ uniformly over $z \in K$.** To see this, let ψ_j, P, Q be as in part **(1)**. Let μ be the peaking function as in part **(2)**. Consider the maps $g_j \equiv \mu \circ \psi_j$. Observe that each g_j is holomorphic and bounded by 1. By Montel's theorem, there is a subsequence g_{j_k} that converges uniformly on compact sets to a limit function g_0.

Of course $|g_0(\zeta)| \leq 1$ for all $\zeta \in U$. But observe that

$$\begin{aligned} g_0(P) &= \lim_{j\to\infty} g_j(P) = \lim_{j\to\infty} \mu(\psi_j(P)) \\ &= \mu\Big(\lim_{j\to\infty} \psi_j(P)\Big) = \mu(Q) = 1. \end{aligned}$$

We have contradicted the maximum modulus principle (refer to [GRK1]) unless $g_0 \equiv 1$. However this means that the ψ_{j_k}s are converging, uniformly on compact sets, to the constant function with value Q. That is what has been claimed.

(4) **There is a small, open disc $\mathcal{D}$ centered at Q such that $U \cap \mathcal{D}$ is simply connected.** In fact this is a straightforward

application of the implicit function theorem (see [KRP1]), and we leave the details to the reader. *Notice that this is the only step in the proof where we use the fact that the boundary curve is continuously differentiable.*

(5) **The domain U is simply connected.** Suppose not. Then we see that there is a closed loop $\gamma : [0,1] \to U$ that cannot be continuously deformed *in* U to a point. However the image curve of γ is a compact set C. Let $\mathcal{D}$ be the disc that we found in step **(4)**. By step **(3)**, there is a k so large that $\psi_{j_k}(C) \subseteq \mathcal{D} \cap U$. But this means that $\psi_{j_k} \circ \gamma$ is a closed curve that lies entirely in the simply connected domain $\mathcal{D} \cap U$. Hence $\psi_{j_k} \circ \gamma$ can certainly be deformed to a point inside $\mathcal{D} \cap U$. But of course the map ψ_j is a homeomorphism. Hence the image C of γ itself may deformed to a point in U. That is a contradiction.

(6) And now our proof is complete, for the simply connected domain U is conformally equivalent to the disc. That is what was claimed. □

We now have a fairly complete understanding of which domains in the complex plane have compact or noncompact automorphism group. In the next section, we shall apply some of our new insights to the study of the dimension of the automorphism group.

We remark that the proof of Theorem 4.4 can be modified to give a new proof of the Riemann Mapping theorem; thus one could, in principle, bypass the use of that theorem in proving Theorem 4.4.

4.3 The Dimension of the Automorphism Group

Capsule: The entire plane is the only complex domain that has automorphism group of dimension four. The disc has automorphism group of dimension three. Most domains have a discrete automorphism group, which has dimension zero.

This section studies how the geometry of a domain affects the dimension of the automorphism group.

If a domain U has very many holes, then the automorphism group will be discrete (i.e., zero-dimensional). If the domain has just one hole, then the group will be one-dimensional. The group can be (at least) two-dimensional if and only if the domain is the disc or

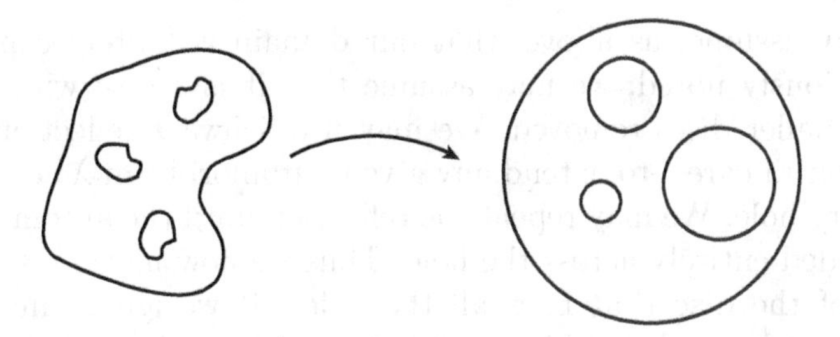

Figure 4.3: A finitely connected domain is equivalent to the disc with finitely many smaller discs removed.

the entire plane or the punctured plane. The purpose of the present section is to flesh out these observations.

The Riemann Mapping theorem says that a simply connected domain (no holes) that is not the entire plane is conformally equivalent to the unit disc. We will make use of a lovely theorem of classical function theory that says that a *finitely connected* domain (i.e., finitely many holes) is conformally equivalent to the unit disc with finitely many smaller discs removed (Figure 4.3). See Sections 1.5 and 1.6.

In fact, He and Schramm [HES] have proved that if a domain has at most *countably many holes*, then it is conformally equivalent to the unit disc with at most countably many smaller discs removed.

If a domain (with C^1 boundary) has automorphism group with infinitely many elements, then $\text{Aut}\,(U)$ is either compact or noncompact. If it is noncompact, then Theorem 4.4 tells us that U is conformally equivalent to the disc. So we may as well assume that the automorphism group is compact. Thus the automorphism group, as a topological space, is compact; so it has an accumulation point. But then the group cannot be a zero-dimensional manifold (i.e., a discrete set). So it must be (at least) a one-dimensional manifold.

Now the only compact one-dimensional manifolds are the circle or a disjoint union of circles. For a domain with finitely many holes, the only way that multiple connected components of the automorphism group can occur is through inversion in one of the holes.[5]

[5]For example, in the simple case that the domain is the annulus $A = \{z \in \mathbb{C} : 1/2 < |z| < 2\}$, the automorphism group is the set of all rotations plus the inversion $z \mapsto 1/z$ plus compositions of the two. Topologically, this automorphism group is two circles. So the automorphism group has two connected components.

Now assume, as above, that our domain is finitely connected. As previously noted, we may assume that it is a disc with finitely many smaller discs removed. We may use Schwarz reflection—with a little extra care—to extend any given automorphism λ across each boundary hole. We may repeat the reflection until the automorphism is extended entirely across the hole. Thus we now have an automorphism of the disc that fixes all the holes. If we ignore inversions, then the only automorphisms that we may consider are those that shuffle the holes. But if the automorphism group is one-dimensional, then there is a one-parameter subgroup φ_t of automorphisms. Fix a circle $\mathcal{C}$ about one of the holes—see Figure 4.4.

Then the mapping

$$t \mapsto \varphi_t(\mathcal{C})$$

describes a continuously parametrized family of closed curves in U. It is impossible (by homotopy-theoretic considerations) for such a family to contain curves that encircle two different holes in the domain. So in fact our one-parameter group of automorphisms must fix all the holes. But then there can be only one hole, because inversion in a hole would create new components of the complement, and the domain is thus (conformally equivalent to) an annulus.

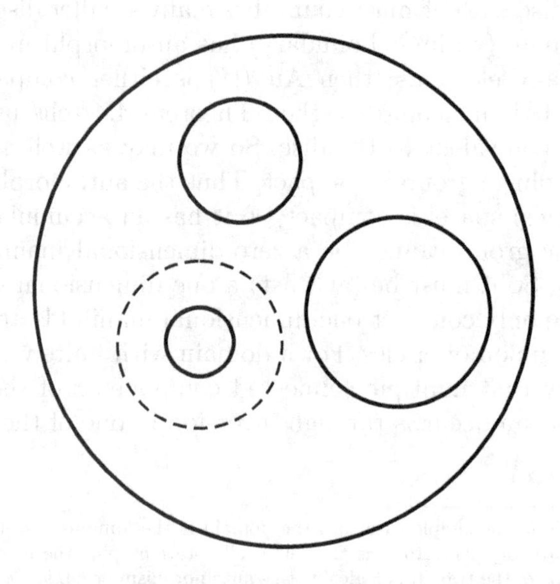

Figure 4.4: A circle about one of the holes.

Our heuristic arguments have shown that:

- The automorphism group of a domain with finitely many holes, but at least two, is finite.
- The only bounded domain with one-dimensional automorphism group is the annulus.

Now which domains have two-dimensional automorphism group? If U is bounded and $\dim(\text{Aut}\,(U)) = 2$, then the orbit of any point $P \in U$ is two-dimensional.[6] It follows then that $\text{Aut}\,(U)$ is noncompact, and hence U is the disc. But we know the automorphism group of the disc explicitly: it is three-dimensional. We conclude that there are no bounded domains with two-dimensional automorphism group. Dimension three is the top dimension possible, and that occurs only when the domain is the disc.

What about unbounded domains? It is easy to see (refer to [GRK1]) that the automorphism group of the entire plane $\mathbb{C}$ is the set of all linear maps

$$z \mapsto az + b,$$

where $a \neq 0$ and b are arbitrary complex constants. Thus $\text{Aut}\,(\mathbb{C})$ has dimension four.

If instead $U = \mathbb{C} \setminus \{0\}$—the punctured plane—then the automorphism group consists of rotations, dilations, and inversion ($z \mapsto 1/z$), and their compositions. Hence the group has dimension two. In fact, one can see that an automorphism of this U either takes the origin to the origin or takes the origin to infinity. In the first case, the point 0 is a "removable singularity" (by Riemann's theorem), and we can see that the mapping is the restriction to U of an automorphism of the plane. This makes it easy to determine the dimension of the group. Recall that we have already observed that there are no *bounded* domains with automorphism group of dimension two.

4.4 The Iwasawa Decomposition

Capsule: The Iwasawa decomposition is a profound idea from Lie theory. For us, in the context of the automorphism group of the disc (or the upper-half-plane), there is a more concrete and practical interpretation.

---

[6]Here the *orbit* of P is defined to be $\{\phi(P) : \phi \in \text{Aut}\,(U)\}$.

The Iwasawa decomposition tells us that the automorphism group of the disc can be factored into the form $K \cdot A \cdot N$, where K is a compact group, A is an abelian group, and N is a nilpotent group.

For the disc, the compact subgroup is the set of automorphisms that fixes the origin. These are just the rotations. The abelian piece is more easily recognized in the context of the upper-half-plane (which of course is conformally equivalent to the disc). There the abelian subgroup is the dilations. Finally, on the upper half-plane, the nilpotent subgroup is the real translations.

We study the Iwasawa decomposition in the present section.

In Lie group theory, there are important structure theorems that help us to understand the specific groups that we study. Sometimes, in concrete situations, these structure theorems take on special significance. We illustrate one of these results in the context of the automorphism group of the unit disc in the complex plane.

First let us formulate a simplified version of the abstract result.

Prelude: This version of the Iwasawa decomposition is quite standard. The result is a fundamental part of modern Lie theory.

THEOREM 4.5 *Let G be a connected, semisimple Lie group. Then we may write*

$$G = K \cdot A \cdot N,$$

where K is a compact Lie group, A is an abelian Lie group, and N is a nilpotent Lie group.

We do not intend to prove the theorem here, but rather to discuss it. A good reference is [HEL]. Now the group we wish to study is $\mathrm{Aut}\,(D)$. For this group, it turns out that

K is the collection of automorphisms that fix the origin.

The subgroups N and A are best understood by looking at the so-called *unbounded realization* of the disc.

That is to say, consider the mapping $\mathcal{C}$ given by

$$\mathcal{C} : D \ni \zeta \mapsto i \cdot \frac{1-\zeta}{1+\zeta} \in \mathcal{U},$$

where $\mathcal{U}$ denotes the upper-half-plane $\mathcal{U} = \{\zeta \in \mathbb{C} : \mathrm{Im}\,\zeta > 0\}$. Then, as usual, we have the topological isomorphism

$$\mathrm{Aut}\,(D) \ni \psi \mapsto \varphi \circ \psi \circ \varphi^{-1} \in \mathrm{Aut}\,(\mathcal{U}).$$

Thus, if we can find the Iwasawa decomposition for $\mathrm{Aut}\,(\mathcal{U})$, then we will have found the decomposition for $\mathrm{Aut}\,(D)$.

You may calculate (exercise) that a linear fractional transformation

$$\varphi : \zeta \mapsto \frac{a\zeta + b}{c\zeta + d}$$

is an automorphism of the upper-half-plane $\mathcal{U}$ if and only if (after normalization by multiplication by a constant in both the numerator and the denominator) a, b, c, and d are real and $ad - bc > 0$.

We let G' denote the automorphism group of $\mathcal{U}$. We write the Iwasawa decomposition as

$$G' = K' \cdot A' \cdot N'.$$

Let $\varphi \in G'$. We wish to decompose

$$\varphi(\zeta) = \frac{a\zeta + b}{c\zeta + d} = \kappa \circ \alpha \circ \nu(\zeta),$$

where $\nu \in N'$ has the form

$$\nu(\zeta) = \zeta + \mu$$

for some real μ; $\alpha \in A'$ has the form

$$\alpha(\zeta) = \lambda\zeta$$

for some positive λ; and $\kappa \in K'$ has the form

$$\kappa(\zeta) = \frac{\gamma\zeta + \delta}{-\delta\zeta + \gamma}$$

for some real γ, δ.

Notice that the elements of N' are left–right translations of $\mathcal{U}$. These form an abelian, hence certainly nilpotent, group.[7] The elements of A' are dilations, thus forming an abelian group. Slightly less obvious is that the elements of K' fix the point $i \in \mathcal{U}$, and hence form a compact group.

It is merely a tedious, but elementary, algebra problem to solve for $\mu, \lambda, \alpha, \beta$. They are uniquely determined, and we find that

$$\nu(\zeta) = \zeta + \left(\frac{dc + ba}{a^2 + c^2}\right),$$

[7] A group is said to be nilpotent if there is an integer K so that any commutator of order K in the group is the identity.

$$\alpha(\zeta) = \frac{a^2 + c^2}{ad - bc} \cdot \zeta,$$

and

$$\kappa(\zeta) = \frac{(a^2d - abc)\zeta + (bc^2 - adc)}{(adc - bc^2)\zeta + (a^2d - abc)}.$$

It is easy to verify that the positivity condition characterizing automorphisms of $\mathcal{U}$ is satisfied by the coefficients of this mapping κ.

We invite you to check that

$$\varphi(\zeta) = \kappa \circ \alpha \circ \nu(\zeta).$$

Now let us translate these three types of mappings back to the disc and see what the Iwasawa decomposition looks like in the original context in which the problem was posed above.

As we have already noted, the compact part of the group is easy. On the upper-half-plane $\mathcal{U}$, this piece K' is the subgroup of automorphisms that fix i. These are mappings of the form

$$\zeta \mapsto \frac{\gamma\zeta + \delta}{-\delta\zeta + \gamma}$$

with γ, δ real. The corresponding subgroup in the disc is the collection of all automorphisms that fix the origin. By the Schwarz lemma, these are just the rotations.

For the nilpotent and abelian pieces we must work a little harder.

Let $\nu' \in N'$. We must calculate $\mathcal{C}^{-1} \circ \nu' \circ \mathcal{C}$ to find the corresponding element of N. First of all,

$$\mathcal{C}^{-1}(\xi) = \frac{i - \xi}{i + \xi}.$$

A typical element of N' is $\xi \mapsto \nu'_\mu(\xi) = \xi + \mu$. Hence a corresponding typical element ν_μ of N will be

$$\begin{aligned} \nu_\mu(\zeta) &= \mathcal{C}^{-1} \circ \nu'_\mu \circ \mathcal{C}(\zeta) \\ &= \frac{i - \left(\frac{i(1-\zeta)}{1+\zeta} + \mu\right)}{i + \left(\frac{i(1-\zeta)}{1+\zeta} + \mu\right)} \\ &= -\left(\frac{\mu - 2i}{\mu + 2i}\right) \cdot \frac{\zeta - \frac{\mu}{-\mu+2i}}{1 - \left(\frac{\mu}{-\mu+2i}\right)\zeta}. \end{aligned}$$

We can plainly see then that $\nu_\mu(\zeta)$ is the composition of a rotation and a Möbius transformation. Notice that $\lim_{\mu\to\pm\infty} \nu_\mu(\zeta) = -1$.

Now let $\alpha' \in A'$. We must calculate $\mathcal{C}^{-1} \circ \alpha' \circ \mathcal{C}$ to find the corresponding element of A. A typical element of A' is $\xi \mapsto \alpha'_\lambda(\xi) = \lambda\xi$ for $\lambda > 0$. Hence a corresponding typical element α_λ of A will be

$$\alpha_\lambda(\zeta) = \mathcal{C}^{-1} \circ \alpha'_\lambda \circ \mathcal{C}(\zeta) = \frac{i - \left(\lambda i \left(\frac{1-\zeta}{1+\zeta}\right)\right)}{i + \left(\lambda i \left(\frac{1-\zeta}{1+\zeta}\right)\right)} = \frac{\zeta - \frac{\lambda-1}{\lambda+1}}{1 - \zeta\left(\frac{\lambda-1}{\lambda+1}\right)}.$$

We see that $\alpha_\lambda(\zeta)$ is a Möbius transformation. Notice that we have the limits $\lim_{\lambda\to 0} \alpha_\lambda(\zeta) = 1$ and $\lim_{\lambda\to+\infty} \alpha_\lambda(\zeta) = -1$.

Example 4.6 Let us consider whether Aut (D) has a subgroup that is group-theoretically isomorphic to the integers $\mathbb{Z}$. With the Iwasawa decomposition of the unbounded realization $\mathcal{U}$, this is obvious. For the translations

$$\nu'_k(\xi) = \xi + k, \quad k \in \mathbb{Z},$$

form a subgroup with the required properties. Notice that this result is not entirely obvious if we work directly with the standard realization of elements of Aut (D) as the composition of rotations with Möbius transformations.

Likewise, we can see that Aut (D) has a subgroup that is isomorphic to the multiplicative group $\mathbb{R}^+$ of positive real numbers. This subgroup, looking at the Iwasawa decomposition of the automorphism group of the unbounded realization $\mathcal{U}$, is just A itself, that is, the dilations

$$\alpha'_\lambda(\xi) = \lambda \cdot \xi.$$

The Iwasawa decomposition can be useful for understanding the orbits of points under the automorphism group action. Fix a point $P \in D$. Under the action of K (the automorphisms that fix 0), the point P is sent to all points of a circle centered at the origin. Under the action of N, the point P is sent to a curve that is the boundary of a horocycle (see [HIL]). Under the action of A, the point P is sent to a geodesic in the Poincaré-Bergman metric. We invite the reader to draw pictures (and to do the attendant calculations) to illustrate these points.

4.5 General Properties of Holomorphic Maps

Capsule: It is natural to wonder whether the algebra of holomorphic functions on a domain (as an algebraic object) characterizes that domain—up to conformal equivalence.

A remarkable and elegant theorem of L. Bers answers this question in the affirmative. We study his result in the present section.

Now we turn to a result of Lipman Bers that has a different flavor. It characterizes domains in terms of a certain algebraic invariant.

Let $U \subseteq \mathbb{C}$ be a domain. Let $\mathcal{O}(U)$ denote the algebra of holomorphic functions from U to $\mathbb{C}$. Bers's theorem says, in effect, that the algebraic structure of $\mathcal{O}(U)$ characterizes U. We begin our study by introducing a little terminology.

Definition 4.7 Let $U \subseteq \mathbb{C}$ be a domain. A $\mathbb{C}$-algebra homomorphism $\varphi : \mathcal{O}(U) \to \mathbb{C}$ is called a *character* of $\mathcal{O}(U)$. If $c \in \mathbb{C}$, then the mapping

$$e_c : \mathcal{O}(U) \to \mathbb{C},$$
$$f \mapsto f(c),$$

is called a *point evaluation*. Every point evaluation is a character.

It should be noted that, if $\varphi : \mathcal{O}(U) \to \mathcal{O}(\widehat{U})$ is not the trivial zero homomorphism, then $\varphi(1) = 1$. This follows because $\varphi(1) = \varphi(1 \cdot 1) = \varphi(1) \cdot \varphi(1)$. On any open set where the holomorphic function $\varphi(1)$ does not vanish, we find that $\varphi(1) \equiv 1$. The result follows by analytic continuation.

It turns out that every character of $\mathcal{O}(U)$ is a point evaluation. That is the content of the next lemma.

LEMMA 4.8 *Let φ be a character on $\mathcal{O}(U)$. Then $\varphi = e_c$ for some $c \in U$. Indeed, $c = \varphi(id) \in U$. Here id is defined by $id(z) = z$.*

Proof: Let c be defined as in the statement of the lemma. Let $f(z) = z - c$. Then

$$\varphi(f) = \varphi(\text{id}) - \varphi(c) = c - c = 0.$$

If it were not the case that $c \in U$, then the function f would be a unit in $\mathcal{O}(U)$. But then

$$1 = \varphi(f \cdot f^{-1}) = \varphi(f) \cdot \varphi(f^{-1}) = 0.$$

That is a contradiction. So $c \in U$.

Now let $g \in \mathcal{O}(U)$ be arbitrary. Then we may write

$$g(z) = g(c) + f(z) \cdot \widetilde{g}(z),$$

where $\widetilde{g} \in \mathcal{O}(U)$. Thus

$$\varphi(g) = \varphi(g(c)) + \varphi(f) \cdot \varphi(\widetilde{g}) = g(c) + 0 = g(c) = e_c(g).$$

We conclude that $\varphi = e_c$, as was claimed. □

Now we may prove Bers's theorem.

Prelude: Bers's theorem establishes that the algebra of holomorphic functions on a domain is a conformal invariant. It has actually been generalized in a variety of ways—to higher dimensions and to more abstract complex structures.

THEOREM 4.9 *Let $U, \widetilde{U}$ be domains. Suppose that*

$$\varphi : \mathcal{O}(U) \to \mathcal{O}(\widetilde{U})$$

is a $\mathbb{C}$-algebra homomorphism. Then there exists one and only one holomorphic mapping $h : \widetilde{U} \to U$ such that

$$\varphi(f) = f \circ h \quad \textit{for all } f \in \mathcal{O}(U).$$

In fact, the mapping h is given by $h = \varphi(\mathrm{id})$.

The homomorphism φ is bijective if and only if h is conformal, that is, h is a one-to-one and onto holomorphic mapping from $\widetilde{U}$ to U.

Proof: Since we want the mapping h to satisfy $\varphi(f) = f \circ h$ for all $f \in \mathcal{O}(U)$, it must, in particular, satisfy $\varphi(\mathrm{id}_U) = \mathrm{id}_U \circ h = h$. We take this as our definition of the mapping h.

If $a \in \widetilde{U}$, then $e_a \circ \varphi$ is a character of $\mathcal{O}(U)$. Thus our lemma tells us that $e_a \circ \varphi$ must be a point evaluation on U. As a result,

$$e_a \circ \varphi = e_c, \quad \text{with } c = (e_a \circ \varphi)(\mathrm{id}_U) = e_a(h) = h(a).$$

Thus, if $f \in \mathcal{O}(U)$, then

$$\varphi(f)(a) = e_a(\varphi \circ f) = (e_a \circ \varphi)(f) = e_{h(a)}(f) = f(h(a)) = (f \circ h)(a)$$

for all $a \in \widetilde{U}$. We conclude that $\varphi(f) = f \circ h$ for all $f \in \mathcal{O}(U)$.

For the last statement of the theorem, suppose that h is a one-to-one, onto conformal mapping of $\widetilde{U}$ to U. If $g \in \mathcal{O}(U)$, then set

$f = g \circ h^{-1}$. It follows that $\varphi(f) = f \circ h = g$. Hence φ is onto. Likewise, if $\varphi(f_1) = \varphi(f_2)$, then $f_1 \circ h = f_2 \circ h$; hence, composing with h^{-1}, $f_1 \equiv f_2$. So φ is one-to-one. Conversely, suppose that φ is an isomorphism. Let $a \in U$ be arbitrary. Then e_a is a character on $\mathcal{O}(U)$; hence $e_a \circ \varphi^{-1}$ is a character on $\mathcal{O}(\widetilde{U})$. By the lemma, there is a point $c \in \widetilde{U}$ such that $e_a \circ \varphi^{-1} = e_c$. It follows that

$$e_a = e_c \circ \varphi.$$

Applying both sides to id_U yields

$$e_a(\mathrm{id}_U) = (e_c \circ \varphi)(\mathrm{id}_U).$$

Unraveling the definitions gives

$$a = e_c(\mathrm{id}_U \circ h) = h(c).$$

Thus $h(c) = a$ and h is surjective. The argument shows that the pre-image c is uniquely determined. So h is also one-to-one. □

As an application of Bers's theorem, we can see immediately that the disc $D = \{z \in \mathbb{C} : |z| < 1\}$ and the annulus $\mathcal{A} = \{z \in \mathbb{C} : 1/2 < |z| < 2\}$ are not conformally equivalent. For the algebra $\mathcal{O}(D)$ can be generated by 1 and z. But the algebra $\mathcal{O}(\mathcal{A})$ cannot be generated by 1 and just one other function (because natural generators for $\mathcal{O}(\mathcal{A})$ are $\{1, z, 1/z\}$ and it is impossible to come up with a shorter list). We leave the details of these assertions to the reader.

Recall from Section 1.6 the concept of essential boundary point. We say that a domain is *maximal* if each boundary point is essential. Given a domain U, there is a unique maximal $U^* \supset U$ (see [FIS, page 66]). Now it is possible to prove the following result.

Theorem: Let U_1, U_2 be two planar domains. Let $\varphi : U_2^* \to U_1^*$ be a one-to-one, onto holomorphic mapping. Then the mapping

$$\begin{aligned} \Phi : H^\infty(U_1) &\to H^\infty(U_2), \\ f &\mapsto f \circ \varphi \end{aligned} \qquad (*)$$

defines a real algebra isomorphism of $H^\infty(U_1)$ onto $H^\infty(U_2)$ with $\Phi(i) = i$ (here i denotes the H^∞ function that is identically equal to i).

Conversely, if Φ is a real algebraic isomorphism of $H^\infty(U_1)$ onto $H^\infty(U_2)$ with $\Phi(i) = i$, then there is a one-to-one, onto mapping $\varphi : U_2 \to U_1$ such that $(*)$ holds.

The proof of this result is not difficult (see [FIS, page 67]). But we shall omit the details and move on to new directions. We now turn to some results about limits of automorphisms. The first was proved by H. Cartan in the 1930s. We begin with some preliminary remarks.

First, if f_j are holomorphic maps from U to U and if f_j converges uniformly on compact sets to a limit function f, and finally if f is not constant, then f is a holomorphic mapping from U to $\overline{U}$. This assertion is immediate from the open mapping principle.

Second, if f_j are holomorphic mappings from U to U, if g_j are holomorphic functions on U, and if

(a) $\lim_{j\to\infty} f_j \equiv f$ is a holomorphic mapping from U to U,

(b) $\lim_{j\to\infty} g_j \equiv g$ is a holomorphic function on U,

then $\lim_{j\to\infty} g_j \circ f_j = g \circ f$. To see this, let $w_j \in U$ be a sequence with limit $w \in U$. Uniform convergence on compact sets gives us that $\lim_{j\to\infty} f_j(w_j) = f(w)$. By the same token, $\lim_{j\to\infty} g_j(f_j(w_j)) = g(f(w))$. We conclude that the sequence $g_j \circ f_j$ converges uniformly on compact sets to $g \circ f$.

Now we turn to Cartan's result.

PROPOSITION 4.10 *Let U be a bounded domain and $\varphi_j \in Aut(U)$. Assume that the φ_j converge uniformly on compact sets to some function f. Then f is either itself an automorphism, or f is constantly equal to some boundary point of U.*

Proof: Of course the limit function f will be holomorphic. Now we examine the sequence $g_j \equiv f_j^{-1}$. These form a bounded sequence of holomorphic functions on U. By Montel's theorem, there is a subsequence g_{j_k} that converges uniformly on compact subsets of U to some element $g \in \mathcal{O}(U)$. We assert that

$$g'(f(w)) \cdot f'(w) \equiv 1 \quad \text{for all } w \in U \text{ with } f(w) \in U. \qquad (4.10.1)$$

To see the assertion, observe that $g_j \circ f_j = \text{id}$ and hence $g_j'(f_j(z)) \cdot f_j'(z) = 1$ for all j and all $z \in U$. Thus we need only check that

$$\lim_{j\to\infty} g_j'(f_j(w)) = g'(f(w)),$$

for all $w \in U \cap f^{-1}(U)$. This assertion holds, however, because $\lim_{j\to\infty} f_j(w) = f(w)$ and the sequence g_j' converges uniformly on compact sets to g'.

With our claim proved, we now note that, if f is not constant, then f is certainly a holomorphic map from U to U (by the first remark preceding the enunciation of our proposition). Also g is not constant by (4.10.1), and again our first remark implies that g is a holomorphic map from U to U. Since $g_j \circ f_j = \text{id} = f_j \circ g_j$, our second remark now implies that

$$g \circ f = \text{id} = f \circ g,$$

hence $f \in \text{Aut}\,(U)$.

But if f is constant then (4.10.1) tells us that $c \equiv f(\Omega)$ cannot be a point in Ω. It follows that $c \in \partial\Omega$. □

A simple, but good, example to keep in mind for illustrating the proposition is $\Omega = D = \{z \in \mathbb{C} : |z| < 1\}$ and

$$f_j(z) = \frac{z - (1 - 1/j)}{1 - (1 - 1/j)z}.$$

Of course each f_j is an automorphism of the disc, but $\lim_{j\to\infty} f_j(z) \equiv -1 \in \partial D$.

The theory becomes particularly rich if we examine not just any sequence of automorphisms but rather the sequence of iterates of a single mapping or automorphism (we have already glimpsed this additional structure in our proof of Cartan's Proposition 4.10). In what follows, if $f : \Omega \to \Omega$ is a holomorphic mapping then we shall use the notation

$$f^j(z) = \underbrace{f \circ f \circ \cdots \circ f}_{j \text{ times}} \quad \text{for } j = 1, 2, \ldots.$$

PROPOSITION 4.11 *Let f be a holomorphic mapping of the domain Ω to itself. Assume that some subsequence f^{j_k} converges uniformly on compact sets to a function $g \in \mathcal{O}(\Omega)$. We conclude that*

(a) *If $g \in Aut\,(\Omega)$ then $f \in Aut\,(\Omega)$.*

(b) *If g is not constant, then every convergent subsequence of the sequence $h_k \equiv f^{j_{k+1}-j_k}$ has the limit function id_Ω.*

Proof: First consider part **(a)**. If it were the case that $f(a) = f(b)$, some $a, b \in \Omega$, then it would follow that $f^j(a) = f^j(b)$ for all j and hence $g(a) = g(b)$. We must conclude that $a = b$ since $g \in \text{Aut}\,(\Omega)$.

For surjectivity, we know that $f^{j_k}(\Omega) \subseteq f(\Omega)$ for every choice of index. It is immediate that $g(\Omega) \subseteq f(\Omega) \subseteq \Omega$, just by set theory.

But we must have that $g(\Omega) = \Omega$ because g is an automorphism. Therefore $f(\Omega) = \Omega$. Assertion **(a)** is now proved.

We turn to **(b)**. By a previous remark, we certainly know that g is a holomorphic mapping from Ω to Ω. Let h be some subsequential limit of the h_k. Then, by an earlier remark, $f_{j_{k+1}} = h_k \circ f_{j_k}$ implies that $g = h \circ g$. Therefore h is the identity on $g(\Omega)$. But $g(\Omega)$ is open in Ω, so we know that $h \equiv \mathrm{id}_\Omega$. That completes the proof. □

One interesting consequence of this last proposition is that, if the iterates f^j of f converge uniformly on compact sets to a nonconstant function, then the limit function is the identity.

Now we have an important result of Cartan.

Prelude: H. Cartan was one of the real pioneers in the study of holomorphic mappings and automorphisms of several complex variables. His theorem is of great utility in studies of automorphism groups.

THEOREM 4.12 (CARTAN) *Let Ω be a bounded domain and let f be a holomorphic mapping of Ω to Ω. Suppose some subsequence f^{j_k} of the iterates of f converges uniformly on compact sets to a nonconstant function g. Then $f \in \mathrm{Aut}\,(\Omega)$.*

Proof: Montel's theorem tells us that $h_k \equiv f^{j_{k+1}-j_k}$ has a subsequence that converges uniformly on compact sets. By part **(b)** of the last proposition, we know that the limit of this subsequence is id_Ω. By part **(a)** of that proposition, $f \in \mathrm{Aut}\,(\Omega)$. □

The next result is philosophically related to the three-fixed-point theorem that we prove below in Section 9.6.

COROLLARY 4.13 *Let Ω be a bounded domain and suppose that $f : \Omega \to \Omega$ is a holomorphic mapping with two distinct fixed points. Then f is an automorphism of Ω.*

Proof: Call the fixed points a and b with $a \neq b$. We assume, of course, that $a \neq b$. Montel's theorem tells us that there is a subsequence $\{f^{j_k}\}$ of the iterates that converges in Ω to a holomorphic function g. Certainly $g(a) = a$ and $g(b) = b$. Hence g is not constant. By the theorem, $f \in \mathrm{Aut}\,(\Omega)$. □

The next corollary is, in part, a reiteration of the theorem of Cartan with which we began this section. Compare it to the classical Schwarz lemma.

COROLLARY 4.14 *Let Ω be a bounded domain and $a \in \Omega$. Then $|f'(a)| \leq 1$ for all holomorphic mappings $f : \Omega \to \Omega$ that fix a. Let $Aut_a(\Omega)$ consist of those automorphisms that fix the point a. Then we have*

$$Aut_a(\Omega) = \{f : f \text{ is a holomorphic mapping from } \Omega \text{ to } \Omega, \\ f(a) = a, \ |f'(a)| = 1\}.$$

Proof: As usual, by Montel, there is a subsequence f^{j_k} of iterates that converges, uniformly on compact sets, to some holomorphic g. One may calculate that

$$\lim_{k\to\infty} [f^{j_k}]'(a) = \lim_{k\to\infty} [f'(a)]^{j_k} = g'(a).$$

But this can be true only if $|f'(a)| \leq 1$ (otherwise the powers of the derivative would blow up).

In case $|f'(a)| = 1$, we see that $|g'(a)| = 1$. But then g is not constant, and Cartan's theorem tells us that $f \in \text{Aut}(\Omega)$.

Conversely, suppose that $f \in \text{Aut}(\Omega)$ and f fixes a. Then $f^{-1} \in \text{Aut}_a(\Omega)$. We know already that $|f'(a)| \leq 1$. Likewise, $|1/f'(a)| = |(f^{-1})'(a)| \leq 1$, or $|f'(a)| \geq 1$. We conclude that $|f'(a)| = 1$. □

We saw some of the interest of $\text{Aut}_a(\Omega)$ in the last corollary. Note that $\text{Aut}_a(\Omega)$ is itself a group, known as the *isotropy subgroup of a*. By Montel's theorem, this subgroup is compact in the topology of uniform convergence on compact sets. Because an automorphism is an isometry in the Bergman metric of Ω, we find it useful to consider the mapping

$$\text{Aut}_a(\Omega) \ni \varphi \mapsto \varphi'(a). \tag{4.15}$$

Applying reasoning as in the last proof, or simply invoking the last corollary, we see that $|\varphi'(a)| = 1$. This mapping is also one-to-one (by reasoning that we have used before), just because an isometry is uniquely determined by where it takes a point (in this case a) and the derivative at that point.

Thus the mapping (4.15) is a one-to-one map of the compact group $\text{Aut}_a(\Omega)$ into the unit circle in the complex plane. The image therefore must be a compact subgroup of the circle group. It is easy to see then that the image is either the entire circle or else a finite subgroup generated by some element of the form $e^{2\pi i/k}$, $k = 1, 2, \ldots$.

An interesting corollary of the reasoning just presented—an interpretation really—is contained in the following result.

PROPOSITION 4.16 *Let Ω be a bounded domain. Fix a point $a \in \Omega$. Let $f \in \mathrm{Aut}_a(\Omega)$. If $f'(a) > 0$, then $f = \mathrm{id}_\Omega$.*

Proof: We know that $f'(a)$ lies on the circle. If $f'(a) > 0$, then $f'(a) = 1$. It follows now from Cartan's theorem that $f = \mathrm{id}_\Omega$. □

Now we bring the argument principle into play to study how the presence of topology adds rigidity to the automorphism group structure.

PROPOSITION 4.17 *Let Ω be a domain. Let f_j be a sequence of functions holomorphic on Ω, and suppose that the sequence converges uniformly on compact sets to a limit function f. Suppose that there is a closed path γ in Ω such that the intersection of the domains interior to $f_j \circ \gamma$ contains at least two points. Then f is nonconstant.*

Proof: Suppose not. If $f(z) \equiv a$, then there would be a point b in the domain interior to the curve $f_j \circ \gamma$, $b \neq a$, such that

$$\frac{1}{2\pi i} \oint_\gamma \frac{f_j'}{f_j - b}\, d\zeta = \frac{1}{2\pi i} \oint_{f_j \circ \gamma} \frac{d\eta}{\eta - b}$$

is a nonzero integer for all but finitely many j. But the sequence $f_j'/(f_j - b)$ converges uniformly to 0 on γ. That certainly gives a contradiction. □

Now our main result about multiply connected domains is as follows.

Prelude: This is an attractive result because it uses the topology of the domain to prevent the mapping from degenerating. Obviously this would make no sense for the disc.

THEOREM 4.18 *Let Ω be a bounded, finitely connected domain in $\mathbb{C}$ with connectivity at least 2. Assume that each connected component of the complement of Ω has at least two points. Let f be a holomorphic self-map of Ω. Assume that every closed path γ in Ω that is not homologous to 0 in Ω has image under f that is also not homologous to 0 in Ω. Then $f \in \mathrm{Aut}(\Omega)$.*

Proof: As usual, we apply Montel's theorem to the sequence of iterates f^j. We find thereby a subsequence f^{j_k} that converges to some limit g. Because Ω has connectivity at least 2 (i.e., the complement has at least two connected components), there is a closed path γ in Ω

that is not homologous to 0. Since $f^j \circ \gamma = f \circ (f^{j-1} \circ \gamma)$, we see that no path $f^{j_k} \circ \gamma$ can be homologous to 0 in Ω. Thus, for each k, there is a connected component K_k of the complement of Ω such that K_k lies in the interior domain of the closed curve $f^{j_k} \circ \gamma$.

Since Ω has just finitely many holes (i.e., bounded connected components of the complement), we may suppose (by passing to a subsequence if necessary and renumbering the holes) that K_k lies in the domain interior to $f^{j_k} \circ \gamma$ for each k. Since K_k has at least two points, we find from our preceding proposition that g is not constant. Now Theorem 4.11 tells us that f is an automorphism of Ω. □

In fact it is known (see [HEI1], [HEI2]) that a domain of finite connectivity at least three will have *finite* automorphism group; that is to say, the domain has only finitely many conformal self-maps. Heins even gave sharp upper bounds for the size of the automorphism group. Let Ω_k have k holes (i.e., Ω_k has connectivity $k+1$). Then the sharp upper bound $N(k)$ for the number of elements in Aut (Ω_k), $k \geq 2$, is

$$\begin{aligned} N(4) &= 12, \\ N(6) &= 24, \\ N(8) &= 24, \\ N(12) &= 60, \\ N(20) &= 60, \\ N(k) &= 2k \text{ for } k \neq 4, 6, 8, 12, 20. \end{aligned}$$

A very interesting open problem is to determine which finite groups arise as the automorphism groups of planar domains (there are some results for finitely connected Riemann surfaces). It is known that, if G is a compact Lie group, then there is some smoothly bounded domain in *some* $\mathbb{C}^n$ with automorphism group equal to G. But it is difficult to say how large n must be in terms of elementary properties of the group G. See [BED], [SAZ], and [GKK] for details.

Chapter 5

The Schwarz Lemma

Prologue: Like normal families, the Schwarz lemma is an elegant and simple idea that turns out to be a powerful tool and a source of much insight and many new ideas.

The lemma was originally conceived as a simple result in complex function theory. But, in the hands of S. T. Yau, L. Ahlfors, H. H. Wu, and many others, it has turned into a powerful geometric tool.

The Schwarz lemma plays a key role in the construction and analysis of the Carathéodory and Kobayashi metrics. It gives a geometric interpretation of the Cauchy estimates and yields important estimates on curvature. And it is a substantial rigidity statement.

Today there are many variants of the Schwarz lemma. There are even Schwarz lemmas at the boundary. The proofs of many of these results are very geometric, and profoundly interesting.

Certainly the Schwarz lemma is a lovely illustration of the richness of mathematical ideas, and of the long reach of very simple ideas. This chapter gives the reader a solid grounding in many aspects of the Schwarz theory.

5.1 Introduction to Schwarz

Capsule: The Schwarz lemma is one of the most simple and elegant results in all of mathematical analysis. But it is also profoundly important. It has influenced many parts of complex function theory, and continues to be studied today.

The classical Schwarz lemma is part of the grist of every complex analysis text. It gives a way to relate the derivative of a holomorphic function to its more global behavior at a lower level. A version of the result is as follows.

LEMMA 5.1 *Let $f : D \to D$ be holomorphic. Assume that $f(0) = 0$. Then*

(a) $|f(z)| \leq |z|$ *for all* $z \in D$.

(b) $|f'(0)| \leq 1$.

At least as important as these two statements are the corresponding uniqueness statements:

(c) *If* $|f(z)| = |z|$ *for some* $z \neq 0$, *then* f *is a rotation:* $f(z) = \lambda z$ *for some unimodular complex constant* λ.

(d) *If* $|f'(0)| = 1$, *then* f *is a rotation:* $f(z) = \lambda z$ *for some unimodular complex constant* λ.

Proof: The classical argument is to consider $g(z) = f(z)/z$. On a circle $|z| = 1 - \epsilon$, we see that $|g(z)| \leq 1/(1-\epsilon)$. Thus $|f(z)| \leq |z|/(1-\epsilon)$. Since this inequality holds for all $\epsilon > 0$, part **(a)** follows. The Cauchy estimates show that $|f'(0)| \leq 1$.

For the uniqueness, if $|f(z)| = |z|$ for some $z \neq 0$, then $|g(z)| = 1$. The maximum modulus principle then forces $|f(z)| = |z|$ for all z, and hence f is a rotation. If, instead, $|f'(0)| = 1$, then $|g(0)| = 1$ and again the maximum modulus principle yields that f is a rotation. □

The Schwarz-Pick lemma takes advantage of the fact that there is no need to restrict to $f(0) = 0$. Once one comes up with the right formulation, the proof is straightforward.

PROPOSITION 5.2 *Let $f : D \to D$. Assume that $a \neq b$ are elements of D and that $f(a) = \alpha$, $f(b) = \beta$. Then*

(a) $\left|\dfrac{\beta - \alpha}{1 - \overline{\alpha}\beta}\right| \leq \left|\dfrac{b - a}{1 - \overline{a}b}\right|$.

(b) $\left|f'(a)\right| \leq \dfrac{1-|\alpha|^2}{1-|a|^2}$.

There is also a pair of uniqueness statements:

(c) *If* $\left|\dfrac{\beta-\alpha}{1-\overline{\alpha}\beta}\right| = \left|\dfrac{b-a}{1-\overline{a}b}\right|$, *then f is a conformal map of the disc D to itself.*

(d) *If* $\left|f'(a)\right| = \dfrac{1-|\alpha|^2}{1-|a|^2}$, *then f is a conformal map of the disc D to itself.*

Proof: We sketch the proof. Recall that, for a a complex number in D,

$$\varphi_a(\zeta) = \frac{\zeta - a}{1-\overline{a}\zeta}$$

defines a *Möbius transformation.* This is a conformal self-map of the disc that takes a to 0. Note that φ_{-a} is the inverse mapping to φ_a.

Now, for the given f, consider

$$g(z) = \varphi_\alpha \circ f \circ \varphi_{-a}.$$

Then $g : D \to D$ and $g(0) = 0$. So the standard Schwarz lemma applies to g. By part **(a)** of that lemma,

$$|g(z)| \leq |z|.$$

Letting $z = \varphi_a(\zeta)$ yields

$$|\varphi_\alpha \circ f(\zeta)| \leq |\varphi_a(\zeta)|.$$

Writing this out, and setting $\zeta = b$, gives the conclusion

$$\left|\frac{\beta-\alpha}{1-\overline{\alpha}\beta}\right| \leq \left|\frac{b-a}{1-\overline{a}b}\right|.$$

That is part **(a)**.

For part **(b)**, we certainly have that

$$\left|(\varphi_\alpha \circ f \circ \varphi_{-a})'(0)\right| \le 1.$$

Using the chain rule, we may rewrite this as

$$\left|\varphi_\alpha'(f \circ \varphi_{-a}(0))\right| \cdot \left|f'(\varphi_{-a}(0))\right| \cdot \left|\varphi_{-a}'(0)\right| \le 1. \qquad (5.2.1)$$

Now

$$\varphi_{-a}'(\zeta) = \frac{1-|a|^2}{(1+\overline{a}\zeta)^2}.$$

So we may rewrite (5.2.1) as

$$\left(\frac{1-|\alpha|^2}{(1-|\alpha|^2)^2}\right) \cdot |f'(a)| \cdot (1-|a|^2) \le 1.$$

Now part **(b)** follows.

We leave parts **(c)** and **(d)** as exercises for the reader. □

The quantity

$$\rho(a,b) = \frac{|a-b|}{|1-\overline{a}b|},$$

which figurs prominently in the statement of the Schwarz-Pick lemma, is called the *pseudohyperbolic metric*. It is actually a metric on D (details left to the reader). It is not identical to the Poincaré-Bergman metric (see Chapter 9 as well as Subsection 5.2.4). In fact, it is not a Riemannian metric at all, but it is still true that conformal maps of the disc are distance-preserving in the pseudohyperbolic metric.

Exercise: Use the Schwarz-Pick lemma to prove this last assertion.

One useful interpretation of the Schwarz-Pick lemma is that a holomorphic function f from the disc to the disc must take each disc $D(0,r)$, $0 < r < 1$, into (but not necessarily onto) the image of that disc under the linear fractional map

$$z \mapsto \frac{z+\alpha}{1+\overline{\alpha}z},$$

where $f(0) = \alpha$. This image is in fact (in case $-1 < \alpha < 1$) a standard Euclidean disc with center on the real axis at α and diameter (in case $0 < \alpha < 1$) given by the interval

$$\left[\frac{\alpha - r}{1-\alpha r}, \frac{\alpha + r}{1+\alpha r}\right].$$

5.2 The Geometry of the Schwarz Lemma

Capsule: It was L. Ahlfors who taught us that the Schwarz lemma can be construed as an inequality about curvature.

In the present section we begin to develop some geometric ideas so that we can appreciate Ahlfors's contribution.

In 1938, Lars Ahlfors [AHL1] caused a sensation by proving that the Schwarz lemma is really an inequality about the curvatures of Riemannian metrics. In the present section and the next we will give an exposition of Ahlfors's ideas. Afterward, we can provide some applications.

We shall go into considerable detail in the present discussion, so that the reader has ample motivation and context.

5.2.1 Metrics

In classical analysis a *metric* is a device for measuring distance. If X is a set, then a metric λ for X is a function

$$\lambda \colon X \times X \longrightarrow \mathbb{R}$$

satisfying, for all $x, y, z \in X$,

(1) $\lambda(x, y) = \lambda(y, x)$

(2) $\lambda(x, y) \geq 0$ and $\lambda(x, y) = 0$ iff $x = y$;

(3) $\lambda(x, y) \leq \lambda(x, z) + \gamma(z, y)$.

The trouble with a metric defined in this generality is that it does not interact well with calculus. In particular, such a metric is not necesarily differentiable. It does not have a natural notion of curvature. It is not obvious how to define and study geodesics. How can we address these failings?

Given two points $p, q \in X$, one would like to consider the *curve of least length* connecting p to q. Any reasonable construction of such a curve leads to a differential equation, and thus we require that our metric lend itself to differentiation. Yet another consideration is curvature: *in the classical setting curvature is measured by the rate of change of the normal vector field.* The concepts of normal and rate of change lead inexorably to differentiation. Thus we shall now take a different approach to the concept of "metric." The reader will see definite parallels here with our treatment of "metric" in Chapter 2 on invariant geometry.

Definition 5.3 If $\Omega \subseteq \mathbb{C}$ is a domain, then a *metric* on Ω is a continuous function $\rho(z) \geq 0$ in Ω that is twice continuously differentiable on $\{z \in \Omega : \rho(z) > 0\}$. If $z \in \Omega$ and $\xi \in \mathbb{C}$ is a tangent vector at z, then we define the *length* of ξ *at* z to be

$$|\xi|_{\rho,z} \equiv \rho(z) \cdot \|\xi\| ,$$

where $\|\xi\|$ denotes the Euclidean length of the vector ξ. We shall use the notation $\| \ \|$ here to denote Euclidean length of a vector.

REMARK 5.4 Usually our metrics ρ will be strictly positive, but it will occasionally be convenient for us to allow a metric to have isolated zeros. These will arise as zeros of holomorphic functions. The zeros of ρ should be thought of as singular points of the metric.

For the record, the metrics we are considering here are a special case of the type of differential metric called *Hermitian*. This terminology need not concern us. Classical analysts sometimes call these metrics *conformal metrics* and write them in the form $\rho(z)|dz|$.

Technically speaking, our metric lives on the tangent bundle to the domain Ω. That is to say, the metric is a function of the variable (z, v), where v is thought of as a tangent vector at the point z. This is just a mathematically rigorous way of saying that the length of the vector v depends on the point z at which it is positioned. The idea of doing geometry in this way goes back to B. Riemann's 1851 doctoral thesis.

Definition 5.5 Let $\Omega \subseteq \mathbb{C}$ be a domain and ρ a metric on U. If

$$\gamma : [a, b] \to \Omega$$

is a continuously differentiable curve then we define its *length in the metric* ρ to be

$$\ell_\rho(\gamma) = \int_a^b |\dot{\gamma}(t)|_{\rho,\gamma(t)}\, dt.$$

The length of a piecewise continuously differentiable curve is defined to be the sum of the lengths of its continuously differentiable pieces.

Classical sources write the arc length as

$$\ell_\rho(\gamma) = \int_\gamma \rho(z)|dz|,$$

but we shall not use this notation.

If a metric ρ is given on a planar domain U, and if P, Q are elements of U, then the distance in the metric ρ from P to Q should be defined as follows: Define $\mathcal{C}_U(P,Q)$ to be the collection of all piecewise continuously differentiable curves $\gamma\colon [0,1] \to U$ such that $\gamma(0) = P$ and $\gamma(1) = Q$. Now define the ρ-metric distance from P to Q to be

$$d_\rho(P,Q) = \inf \{\ell_\rho(\gamma)\colon \gamma \in \mathcal{C}_U(P,Q)\}.$$

Check for yourself that the resulting notion of distance satisfies the classical metric axioms listed at the outset of this section.

REMARK 5.6 There is some subtlety connected with defining distance in this fashion. If $\rho(z) \equiv 1$, the Euclidean metric, and if U is the entire plane, then $d_\rho(P,Q)$ is the ordinary Euclidean distance from P to Q. The "shortest curve" from P to Q in this setting is the usual straight line segment. But if U and P and Q are as shown in Figure 5.1, then there is no shortest curve in U connecting P to Q. The *distance* from P to Q is suggested by the dotted curve, but notice that this curve does not lie in U. The crucial issue here

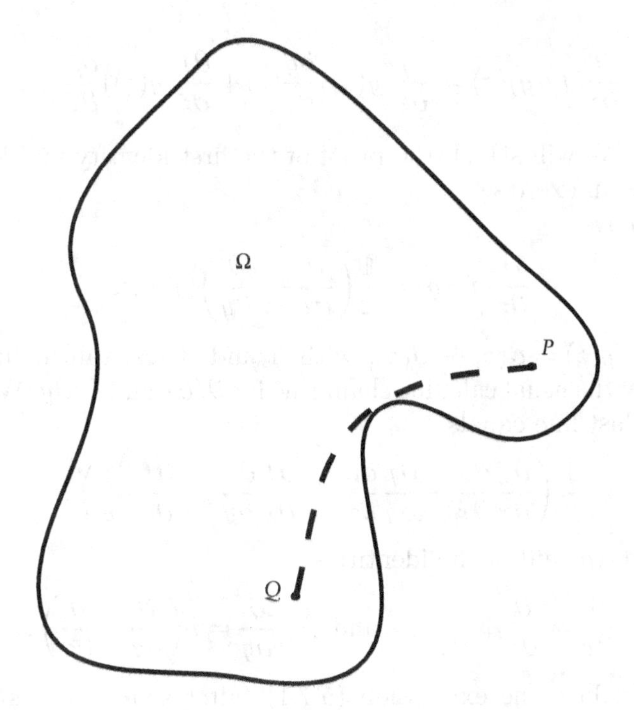

Figure 5.1: No shortest curve.

is whether the domain is complete in the metric, and we shall have more to say about this point later on.

5.2.2 Calculus from the Complex Viewpoint

In order that we may be able to do calculus computations easily and efficiently in the context of complex analysis, we recast some of the basic ideas in new notation. We define the differential operators

$$\frac{\partial}{\partial z} = \frac{1}{2}\left(\frac{\partial}{\partial x} - i\frac{\partial}{\partial y}\right) \quad \text{and} \quad \frac{\partial}{\partial \bar{z}} = \frac{1}{2}\left(\frac{\partial}{\partial x} + i\frac{\partial}{\partial y}\right).$$

This is, in effect, a new basis for the tangent space to $\mathbb{C}$. In complex analysis, it is more convenient to use these operators than to use $\partial/\partial x$ and $\partial/\partial y$.

PROPOSITION 5.7 *If f and g are continuously differentiable functions, and if $f \circ g$ is well defined on some open set $U \subseteq \mathbb{C}$, then we have*

$$\frac{\partial}{\partial z}(f \circ g)(z) = \frac{\partial f}{\partial z}(g(z))\frac{\partial g}{\partial z}(z) + \frac{\partial f}{\partial \bar{z}}(g(z))\frac{\partial \bar{g}}{\partial z}(z)$$

and

$$\frac{\partial}{\partial \bar{z}}(f \circ g)(z) = \frac{\partial f}{\partial z}(g(z))\frac{\partial g}{\partial \bar{z}}(z) + \frac{\partial f}{\partial \bar{z}}(g(z))\frac{\partial \bar{g}}{\partial \bar{z}}(z).$$

Proof: We will sketch the proof of the first identity and leave the second as an exercise.

We have

$$\frac{\partial}{\partial z}(f \circ g) = \frac{1}{2}\left(\frac{\partial}{\partial x} - i\frac{\partial}{\partial y}\right)(f \circ g).$$

We write $g(z) = \alpha(z) + i\beta(z)$, with α and β real-valued functions, and apply the usual calculus chain rule for $\partial/\partial x$ and $\partial/\partial y$. We obtain that the last line equals

$$\frac{1}{2}\left(\frac{\partial f}{\partial x}\frac{\partial \alpha}{\partial x} + \frac{\partial f}{\partial y}\frac{\partial \beta}{\partial x} - i\frac{\partial f}{\partial x}\frac{\partial \alpha}{\partial y} - i\frac{\partial f}{\partial y}\frac{\partial \beta}{\partial y}\right). \tag{5.7.1}$$

Now, with the aid of the identities

$$\frac{\partial}{\partial x} = \frac{\partial}{\partial z} + \frac{\partial}{\partial \bar{z}} \quad \text{and} \quad \frac{\partial}{\partial y} = i\left(\frac{\partial}{\partial z} - \frac{\partial}{\partial \bar{z}}\right),$$

we may reduce the expression (5.7.1) (after some tedious calculations) to the desired formula. □

COROLLARY 5.8 *If either f or g is holomorphic, then*

$$\frac{\partial}{\partial z}(f \circ g)(z) = \frac{\partial f}{\partial z}(g(z))\frac{\partial g}{\partial z}(z).$$

Here is an example of the utility of our complex calculus notation.

PROPOSITION 5.9 *Let f be a nonvanishing holomorphic function on a planar domain U. Then*

$$\Delta\big(\log(\|f\|^2)\big) = 0.$$

In other words, $\log(\|f\|^2)$ is harmonic.

Proof: Fix $P \in U$ and let $V \subseteq U$ be a neighborhood of P on which f has a holomorphic logarithm. Then, on V, we have

$$\begin{aligned}\Delta\big(\log(\|f\|^2)\big) &= \Delta\big(\log f + \log \overline{f}\big)\\ &= 4\frac{\partial}{\partial z}\frac{\partial}{\partial \overline{z}}\log f + 4\frac{\partial}{\partial \overline{z}}\frac{\partial}{\partial z}\log \overline{f}\\ &= 0.\end{aligned}$$

Of course the reader may also check that $\log\|f\|$ is harmonic. □

We conclude this section with some exercises for the reader:

Exercises

1. Calculate that, for $a > 0$,

$$\frac{\partial^2}{\partial z \partial \overline{z}} \log\Big(1 + (z\overline{z})^a\Big) = \frac{a^2 (z\overline{z})^{a-1}}{(1+(z\overline{z})^a)^2}.$$

2. If g is holomorphic (and f continuously differentiable), then calculate that

$$\Delta(f \circ g) = (\Delta f \circ g) \cdot |g'|^2.$$

3. If f is holomorphic (and g continuously differentiable), then calculate that

$$\Delta(f \circ g) = (f' \circ g)\,\Delta\, g + (f'' \circ g)[(D_x g)^2 + (D_y g)^2].$$

5.2.3 Isometries

In any mathematical subject there are morphisms: functions that preserve the relevant properties being studied. In linear algebra these are linear maps; in Euclidean geometry these are rigid motions; and in Riemannian geometry these are "isometries." We now define the concept of isometry.

Definition 5.10 Let U_1 and U_2 be planar domains and let

$$f : U_1 \to U_2$$

be a continuously differentiable mapping with Jacobian having isolated zeros. Assume that U_2 is equipped with a metric ρ. We define the *pullback* of the metric ρ under the map f to be the metric on U_1 given by

$$f^*\rho(z) = \rho(f(z)) \cdot \left\| \frac{\partial f}{\partial z} \right\| .$$

REMARK 5.11 The particular form that we use to define the pullback is motivated by the way that f induces mappings on tangent and cotangent vectors, but this motivation is irrelevant for us here.

It should be noted that the pullback of any metric under a conjugate holomorphic f will be the zero metric. Thus we have designed our definition of pullback so that holomorphic pullbacks will be the ones of greatest interest. This assertion will be made substantive in Propositions 5.13 and 5.14 below.

Definition 5.12 Let U_1, U_2 be planar domains equipped with metrics ρ_1 and ρ_2, respectively. Let

$$f : U_1 \to U_2$$

be an onto, holomorphic mapping. If

$$f^*\rho_2(z) = \rho_1(z)$$

for all $z \in U_1$, then f is called an *isometry* of the pair (U_1, ρ_1) with the pair (U_2, ρ_2).

The differential definition of isometry is very natural from the point of view of differential geometry, but it is not necessarily intuitive. You can check any differential geometry text and see that what we want to do is to map the tangent spaces in a canonical fashion. The next proposition relates the notion of isometry to more familiar geometric ideas.

PROPOSITION 5.13 *Let U_1, U_2 be domains and ρ_1, ρ_2 be metrics on these respective domains. If*

$$f : U_1 \to U_2$$

is a holomorphic isometry (in particular, an onto mapping) of (U_1, ρ_1) to (U_2, ρ_2), then the following three properties hold:

(a) *If $\gamma : [a, b] \to U_1$ is a continuously differentiable curve then so is the push-forward $f_*\gamma \equiv f \circ \gamma$ and*

$$\ell_{\rho_1}(\gamma) = \ell_{\rho_2}(f_*\gamma).$$

(b) *If P, Q are elements of U_1, then*

$$d_{\rho_1}(P, Q) = d_{\rho_2}(f(P), f(Q)).$$

(c) *Part* **(b)** *implies that the isometry f is one-to-one. Then f^{-1} is well defined and f^{-1} is also an isometry.*

Proof: Assertion **(b)** is an immediate consequence of **(a)**. Also **(c)** is a formal exercise in definition chasing. Therefore we shall prove **(a)**.

By definition,

$$\ell_{\rho_2}(f_*\gamma) = \int_a^b \left|(f_*\gamma)'(t)\right|_{\rho_2, f_*\gamma(t)} dt = \int_a^b \left|\frac{\partial f}{\partial z}(\gamma(t)) \cdot \dot{\gamma}(t)\right|_{\rho_2, f_*\gamma(t)} dt.$$

With elementary manipulations, we see that the integrand equals

$$\begin{aligned} \left\|\frac{\partial f}{\partial z}(\gamma(t))\right\| \cdot |\dot{\gamma}(t)|_{\rho_2, f_*\gamma(t)} &= \left\|\frac{\partial f}{\partial z}(\gamma(t))\right\| \cdot \|\dot{\gamma}(t)\| \cdot \rho_2(f(\gamma(t))) \\ &= |\dot{\gamma}(t)|_{f^*\rho_2, \gamma(t)} \\ &= |\dot{\gamma}(t)|_{\rho_1, \gamma(t)}, \end{aligned}$$

since f is an isometry. Substituting this result back into the formula for the length of $f_*\gamma$ gives

$$\ell_{\rho_2}(f_*\gamma) = \int_a^b |\dot{\gamma}(t)|_{\rho_1, \gamma(t)} dt = \ell_{\rho_1}(\gamma).$$

This ends the proof. □

In the next section, we shall consider the isometries of the Poincaré metric on the disc.

If you have previously studied metric space theory or Banach space theory, then you may have already encountered the term "isometry." The essential notion is that an isometry should preserve distance. In fact you can prove as an exercise that if f is a holomorphic mapping of (U_1, ρ_1) onto (U_2, ρ_2) that preserves distance, then f is an isometry according to Definition 5.12. (Hint: compose f with a curve and differentiate.)

We close this section by noting an important technical fact about isometries:

PROPOSITION 5.14 *Let ρ_j be metrics on the domains U_j, $j = 1, 2, 3$, respectively. Let $f : U_1 \to U_2$ and $g : U_2 \to U_3$ be isometries. Then $g \circ f$ is an isometry of the metric ρ_1 to the metric ρ_3.*

Proof: We calculate that

$$\rho_3(g([f(z)])) \cdot \|g'(f(z))\| = \rho_2(f(z))$$

hence

$$\rho_3(g([f(z)])) \cdot \|g'(f(z))\| \cdot \|f'(z)\| = \rho_2(f(z)) \cdot \|f'(z)\| = \rho_1(z).$$

In other words,

$$\rho_3(g \circ f(z)) \cdot \|(g \circ f)'(z)\| = \rho_1(z),$$

as required by the definition of an isometry. □

5.2.4 The Poincaré Metric

The Poincaré metric on the disc has considerable historic significance. This metric is the paradigm for much of what we want to do in the present chapter, and we shall treat it in some detail here. The Poincaré metric on the disc D is given by

$$\rho(z) = \frac{1}{1 - |z|^2}.$$

(For the record, we note that there is no agreement in the literature as to what constant goes in the numerator; many references use a factor of 2.)

In this and succeeding sections, we shall use the phrase "conformal map" to refer to a holomorphic mapping of one planar domain to another that is both one-to-one and onto (and thus invertible).

PROPOSITION 5.15 *Let ρ be the Poincaré metric on the disc D. Let $h : D \to D$ be a conformal self-map of the disc. Then h is an isometry of the pair (D, ρ) with the pair (D, ρ).*

Proof: We have that

$$h^*\rho(z) = \rho(h(z)) \cdot \|h'(z)\|.$$

We now have two cases:

(i) If h is a rotation, then $h(z) = \mu \cdot z$ for some unimodular constant $\mu \in \mathbb{C}$. So $\|h'(z)\| = 1$ and

$$h^*\rho(z) = \rho(h(z)) = \rho(\mu z) = \frac{1}{1 - \|\mu z\|^2} = \frac{1}{1 - \|z\|^2} = \rho(z)$$

as desired.

(ii) If h is a Möbius transformation, then

$$h(z) = \frac{z-a}{1-\overline{a}z}, \qquad \text{some constant } a \in D.$$

But then

$$\|h'(z)\| = \frac{1 - \|a\|^2}{\|1 - \overline{a}z\|^2}$$

and

$$\begin{aligned} h^*\rho(z) &= \rho\left(\frac{z-a}{1-\overline{a}z}\right) \cdot \|h'(z)\| \\ &= \frac{1}{1 - \left\|\frac{z-a}{1-\overline{a}z}\right\|^2} \cdot \frac{1-\|a\|^2}{\|1-\overline{a}z\|^2} \\ &= \frac{1-\|a\|^2}{\|1-\overline{a}z\|^2 - \|z-a\|^2} \\ &= \frac{1-\|a\|^2}{1 - \|z\|^2 - \|a\|^2 + \|a\|^2\|z\|^2} \\ &= \frac{1}{1-\|z\|^2} \\ &= \rho(z). \end{aligned}$$

Since any conformal self-map of D is a composition of maps of the form **(i)** or **(ii)**, the proposition is proved. □

REMARK 5.16 Certainly the last proposition can be derived from the Schwarz-Pick lemma. Do this as an exercise.

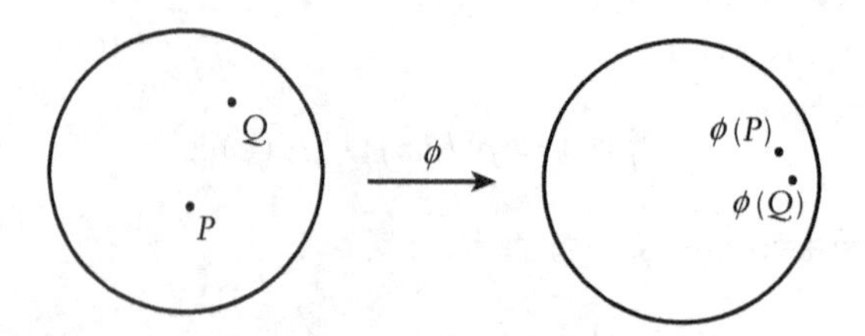

Figure 5.2: Map preserves Poincaré distance but not Euclidean distance.

We know from this last result that conformal self-maps of the disc preserve Poincaré distance. To understand what this means, consider the Möbius transformation

$$\varphi(z) = \frac{z+a}{1+\overline{a}z}.$$

Then φ maps the disc to the disc conformally. It does not preserve Euclidean distance (for instance, look at the images of 0 and 1/2), but it *does* preserve Poincaré distance. See Figure 5.2.

We can use what we have learned so far to calculate the Poincaré metric explicitly.

PROPOSITION 5.17 *If P and Q are points of the disc D, then the Poincaré distance of P to Q is*

$$d_\rho(P,Q) = \frac{1}{2}\log\left(\frac{1+\left\|\frac{P-Q}{1-\overline{P}Q}\right\|}{1-\left\|\frac{P-Q}{1-\overline{P}Q}\right\|}\right).$$

Proof: In case $P = 0$ and $Q = R + i0$, then we simply note (as a simple estimate shows) that the shortest path between these points is a straight line segment (see [KRA1] for the details). Then one can explicitly calculate the Poincaré distance.

In the general case, note that we may define

$$\varphi(z) = \frac{z-P}{1-\overline{P}z},$$

a Möbius transformation of the disc. Then

$$d_\rho(P,Q) = d_\rho(\varphi(P), \varphi(Q)) = d_\rho(0, \varphi(Q)).$$

Next we have

$$d_\rho(0, \varphi(Q)) = d_\rho(0, \|\varphi(Q)\|) \qquad (5.17.1)$$

since there is a rotation of the disc taking $\varphi(Q)$ to $\|\varphi(Q)\| + i0$. Finally,

$$|\varphi(Q)| = \left\| \frac{P-Q}{1-\overline{P}Q} \right\|,$$

so that (5.17.1) together with the special case treated in the the first sentence gives the result. □

Notice the pseudohyperbolic metric rearing its head again here. We first encountered that metric in Section 5.1.

One of the reasons that the Poincaré metric is so useful is that it induces the same topology as the usual Euclidean metric topology. That is our next result. In fact a similar result holds for any Riemannian metric—just because we construct the new metric by integrating a positive function.

PROPOSITION 5.18 *The topology induced on the disc by the Poincaré metric is the usual planar topology.*

Proof: A neighborhood basis for the topology of the Poincaré metric at the origin is given by the balls

$$\mathbf{B}(0, r) = \{z : d_\rho(0, z) < r\}.$$

However, a calculation using Proposition 5.17 yields that these balls are the same as the Euclidean discs

$$\left\{ z : \|z\| < \frac{e^{2r} - 1}{e^{2r} + 1} \right\}.$$

These discs form a neighborhood basis for the origin in the Euclidean topology. Thus we find that the two topologies are the same at the origin. Now the origin can be moved to any other point $a \in D$ by the Möbius transformation

$$z \mapsto \frac{z + a}{1 + \overline{a}z}.$$

Since the Poincaré metric is invariant under Möbius transformations, and since Möbius transformations take circles to circles (after all, they are linear fractional) and hence discs to discs, the two topologies are the same at every point. □

One of the most striking facts about the Poincaré metric on the disc is that it turns the disc into a *complete* metric space. How could this be? The boundary is missing! The reason that the disc is complete in the Poincaré metric is the same as the reason that the plane is complete in the Euclidean metric: the boundary is infinitely far away. We now prove this assertion.

PROPOSITION 5.19 *The unit disc D, when equipped with the Poincaré metric, is a complete metric space.*

Proof: Let p_j be a sequence in D that is Cauchy in the Poincaré metric. Then the sequence is bounded in that metric. So there is a positive, finite M such that

$$d_\rho(0, p_j) \leq M, \qquad \text{all } j.$$

With Proposition 5.17 this translates to

$$\frac{1}{2}\log\left(\frac{1+|p_j|}{1-|p_j|}\right) \leq M.$$

Solving for $|p_j|$ gives

$$|p_j| \leq \frac{e^{2M}-1}{e^{2M}+1} < 1.$$

Thus our sequence is contained in a relatively compact subset of the disc. A similar calculation yields that the sequence must, in fact, be Cauchy in the Euclidean metric. Therefore it converges to a limit point in the disc, as required for completeness. □

In the proof of Proposition 5.18 and in the subsequent remarks, we used implicitly the fact, whose verification was sketched earlier, that the curve of least length (in the Poincaré metric) connecting 0 to a point of the form $R + i0$ is a Euclidean segment. More generally, the shortest path from 0 to any point w is a rotation of the shortest path from 0 to $\|w\| + i0$, which is a segment. Let us now calculate the "curve of least length" connecting any two given points P and Q in the disc.

PROPOSITION 5.20 *Let P, Q be elements of the unit disc. The "curve of least Poincaré length" connecting P to Q is*

$$\gamma_{P,Q}(t) = \frac{t\frac{Q-P}{1-Q\overline{P}} + P}{1 + t\overline{P}\cdot\frac{Q-P}{1-Q\overline{P}}}, \qquad 0 \leq t \leq 1.$$

Proof: Define the Möbius transformation

$$\varphi(z) = \frac{z-P}{1-\overline{P}z}.$$

By what we already know about shortest paths emanating from the origin, the curve $\tau(t) \equiv t \cdot \varphi(Q)$ is the shortest curve from $\varphi(P) = 0$ to $\varphi(Q)$, $0 \le t \le 1$. Applying the isometry

$$\varphi^{-1}(z) = \frac{z+P}{1+\overline{P}z}$$

to τ, we obtain that

$$\varphi^{-1} \circ \tau(t) = \frac{(t\varphi(Q)+P)}{(1+\overline{P}t\varphi(Q))}$$

is the shortest path from P to Q. Since $\varphi^{-1} \circ \tau = \gamma_{P,Q}$, we are done. □

Let us analyze the curves discovered in Proposition 5.20. First notice that, since the curve $\gamma_{P,Q}$ is the image under a linear fractional transformation of a part of a line, the trace of $\gamma_{P,Q}$ is, therefore, either a line segment or an arc of a circle. In fact, if P and Q are collinear with 0, then the formula for $\gamma_{P,Q}$ quickly reduces to that for a segment; otherwise $\gamma_{P,Q}$ traces an arc of a Euclidean circle. Which circle is it?

Matters are simplest if we let t range over the entire real line and look for the whole circle. We find that $t = \infty$ corresponds to the point $1/\overline{P}$. It is now a simple, but tedious, calculation to determine the Euclidean center and radius of the circle determined by the three points $P, Q, 1/\overline{P}$ (note that, by symmetry, the circle will also pass through $1/\overline{Q}$). The circle is depicted in Figure 5.3. Notice that, since the segment $\{t+i0 : -1 \le t \le 1\}$ is orthogonal to ∂D at the endpoints 1 and -1, conformality dictates that the circular arcs of least length provided by Proposition 5.20 are orthogonal to ∂D at the points of intersection. Some of these "geodesic arcs" are exhibited in Figure 5.4.

A final note on this matter is that geodesics are particularly easy to calculate in the upper-half-plane realization $\mathcal{U}$ of the disc. Here $\mathcal{U} = \{x+iy : y > 0\}$. For the geodesic circles will then have their centers on the boundary (i.e., the real line). If $\widetilde{P}, \widetilde{Q}$ are points of $\mathcal{U}$—not both on the same vertical line—then the perpendicular bisector of the segment connecting these points will intersect the real axis at

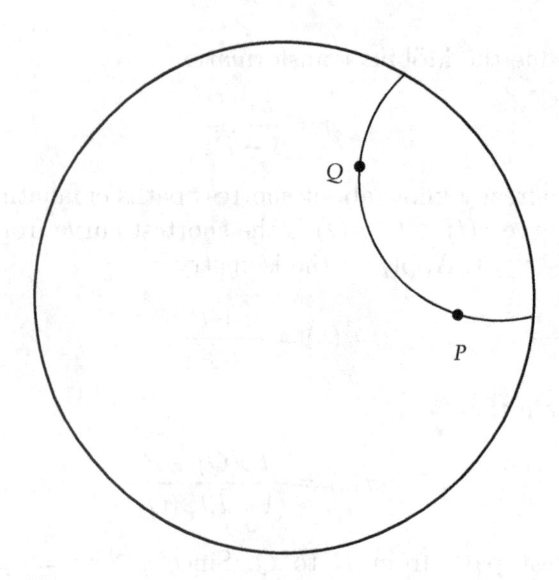

Figure 5.3: Calculation of geodesic arcs.

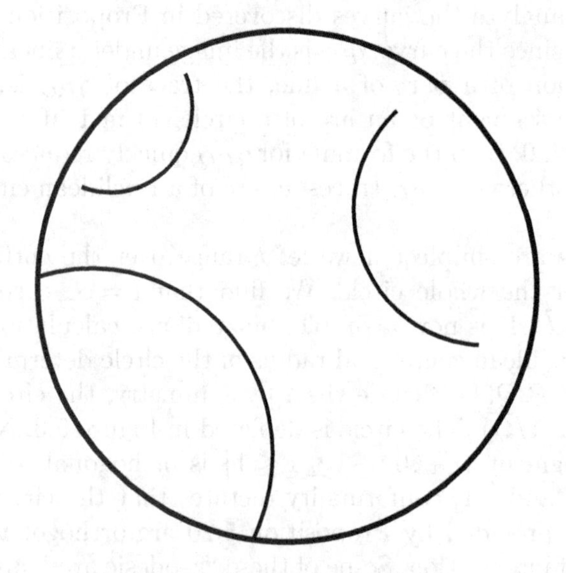

Figure 5.4: Some geodesic arcs.

the center C of the circular arc that forms the geodesic through $\widetilde{P}$ and $\widetilde{Q}$. See Figure 5.5.

We next see that the Poincaré metric is characterized by its property of invariance under conformal maps. So, in a sense, the Poincaré metric is the *unique* invariant metric on the unit disc.

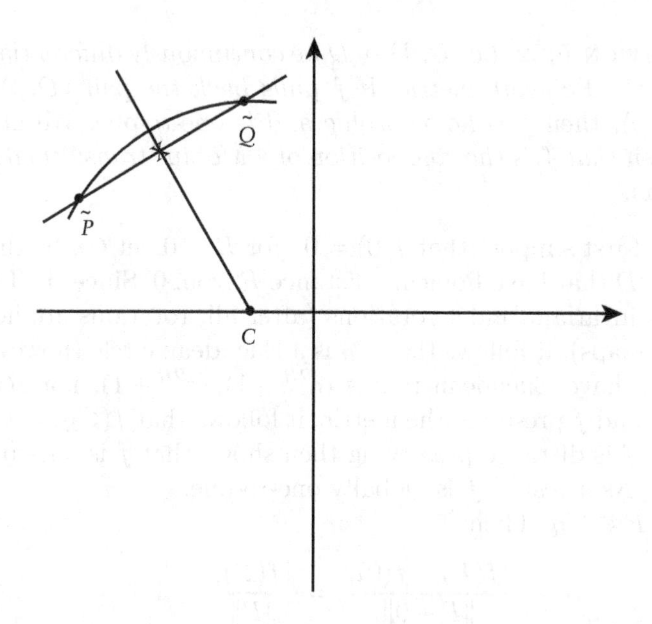

Figure 5.5: A Poincaré geodesic in the upper half-plane.

PROPOSITION 5.21 *If $\widetilde{\rho}(z)$ is a metric on D that is such that every conformal map of the disc is an isometry of the pair $(D, \widetilde{\rho})$ with the pair $(D, \widetilde{\rho})$ then $\widetilde{\rho}$ is a constant multiple of the Poincaré metric ρ.*

Proof: The hypothesis guarantees that, if $z_0 \in D$ is fixed and

$$h(z) = \frac{z + z_0}{1 + \overline{z}_0 z},$$

then

$$h^* \widetilde{\rho}(0) = \widetilde{\rho}(0).$$

Writing out the left-hand side gives

$$\|h'(0)\| \widetilde{\rho}(h(0)) = \widetilde{\rho}(0),$$

or

$$\widetilde{\rho}(z_0) = \frac{1}{1 - \|z_0\|^2} \cdot \widetilde{\rho}(0) = \widetilde{\rho}(0) \cdot \rho(z_0).$$

Thus we have exhibited $\widetilde{\rho}$ as the constant $\widetilde{\rho}(0)$ times ρ. □

Now that we know that the Poincaré metric is the right metric for complex analysis on the disc, a natural next question is to determine which other maps preserve the Poincaré metric.

PROPOSITION 5.22 *Let $f : D \to D$ be continuously differentiable and let ρ be the Poincaré metric. If f pulls back the pair (D, ρ) to the pair (D, ρ), then f is holomorphic and is one-to-one. We may conclude then that f is the composition of a Möbius transformation and a rotation.*

Proof: First suppose that $f(0) = 0$. For $R > 0$, let C_R be the set of points in D that have Poincaré distance R from 0. Since the Poincaré metric is invariant under rotations (after all, rotations are holomorphic self-maps), it follows that C_R is a Euclidean circle (however, this circle will have Euclidean radius $(e^{2R} - 1)/(e^{2R} + 1)$, not R). Since $f(0) = 0$ and f preserves the metric, it follows that $f(C_R) = C_R$. The fact that f is distance-preserving then shows that f is one-to-one on each C_R. As a result, f is globally one-to-one.

Let $P \in C_R$. Then

$$\frac{\|f(P) - f(0)\|}{\|P - 0\|} = \frac{\|f(P)\|}{\|P\|} = 1.$$

Letting $R \to 0^+$, we conclude that f is conformal at the origin (that is, f preserves length infinitesimally). Since f pulls back the metric ρ, we can be sure that $\partial\rho/\partial z \neq 0$ at the origin.

Now we drop the special hypothesis that $f(0) = 0$. Pick an arbitrary $z_0 \in D$, and set $w_0 = f(z_0)$. Define

$$\varphi(z) = \frac{z + z_0}{1 + \overline{z}_0 z}, \qquad \psi(z) = \frac{z - w_0}{1 - \overline{w}_0 z}.$$

Also define

$$g = \psi \circ f \circ \varphi.$$

Then $g(0) = 0$ and g is an isometry (since ψ, f, and φ are); therefore the argument in the last paragraph applies and g is conformal at the origin. It follows as before that $\partial g/\partial z \neq 0$ at the origin.

We conclude that f is conformal at every point—since ψ and ϕ are—and $\partial f/\partial z \neq 0$ at every point. Therefore f is holomorphic. □

Isometries are very rigid objects. They are completely determined by their first-order behavior at just one point. While a proof of this assertion in general is beyond us at this point, we can certainly prove the result for the Poincaré metric on the disc.

PROPOSITION 5.23 *Let ρ be the Poincaré metric on the disc. Let f be an isometry of the pair (D, ρ) with the pair (D, ρ). If $f(0) = 0$ and $\partial f/\partial z(0) = 1$, then $f(z) \equiv z$.*

Proof: By Proposition 5.22, f must be holomorphic. Since f preserves the origin and is one-to-one and onto, f must be a rotation. Since $f'(0) = 1$, it follows that f is the identity. □

COROLLARY 5.24 *Let f and g be isometries of the pair (D, ρ) with the pair (D, ρ). Let $z_0 \in D$ and suppose that $f(z_0) = g(z_0)$ and $(\partial f/\partial z)(z_0) = (\partial g/\partial z)(z_0)$. Then $f(z) \equiv g(z)$.*

Proof: We noted earlier in this section that g^{-1} is an isometry. If ψ is a Möbius transformation that takes 0 to z_0, then $\psi^{-1} \circ g^{-1} \circ f \circ \psi$ satisfies the hypothesis of the proposition. As a result, $\psi^{-1} \circ g^{-1} \circ f \circ \psi(z) \equiv z$ or $g(z) \equiv f(z)$. □

5.2.5 The Spirit of the Schwarz Lemma

One of the important facts about the Poincaré metric is that it can be used to study not just conformal maps but all holomorphic maps of the disc. The key to this assertion is the classical Schwarz lemma. We begin with an elegant geometric interpretation of the Schwarz-Pick lemma.

PROPOSITION 5.25 *Let $f : D \to D$ be holomorphic. Then f is distance-decreasing in the Poincaré metric ρ. That is, for any $z \in D$,*

$$f^*\rho(z) \le \rho(z).$$

The integrated form of this assertion is that, if $\gamma : [0, 1] \to D$ is a continuously differentiable curve, then

$$\ell_\rho(f_*\gamma) \le \ell_\rho(\gamma).$$

Therefore, if P and Q are elements of D, we may conclude that

$$d_\rho(f(P), f(Q)) \le d_\rho(P, Q).$$

Proof: Now

$$f^*\rho(z) \equiv \|f'(z)\|\rho(f(z)) = \|f'(z)\| \cdot \frac{1}{1 - \|f(z)\|^2}$$

and

$$\rho(z) = \frac{1}{1 - \|z\|^2},$$

so the asserted inequality is just the Schwarz-Pick lemma. The integrated form of the inequality now follows by the definition of ℓ_ρ.

The inequality for the distance d_ρ follows now directly from the definition of distance. □

Notice that, in case f is a conformal self-map of the disc, we may apply this corollary to both f and f^{-1} to conclude that f preserves Poincaré distance, giving a second proof of Proposition 5.15.

We next give an illustration of the utility of the geometric point of view. We will see that the proposition just proved gives a very elegant proof of Theorem 5.26 below (see [EAH] for the source of this proof).

Prelude: One should compare this theorem with the Brouwer fixed-point theorem. For that topological result, it is essential to deal with compact spaces. What is interesting about the Farkas/Ritt result is that it is about open domains. The geometry is what is essential here.

THEOREM 5.26 (FARKAS, RITT) *Let $f : D \to D$ be holomorphic and assume that the image $M = \{f(z) : z \in D\}$ of f has compact closure in D. Then there is a unique point $P \in D$ such that $f(P) = P$. We call P a fixed point for f.*

Proof: By hypothesis, there is an $\epsilon > 0$ such that if $m \in M$ and $\|z\| \geq 1$, then $\|m - z\| > 2\epsilon$. See Figure 5.6. Fix $z_0 \in D$ and define

$$g(z) = f(z) + \epsilon(f(z) - f(z_0)).$$

Then g is holomorphic and g still maps D into D. Also

$$g'(z_0) = (1 + \epsilon)f'(z_0).$$

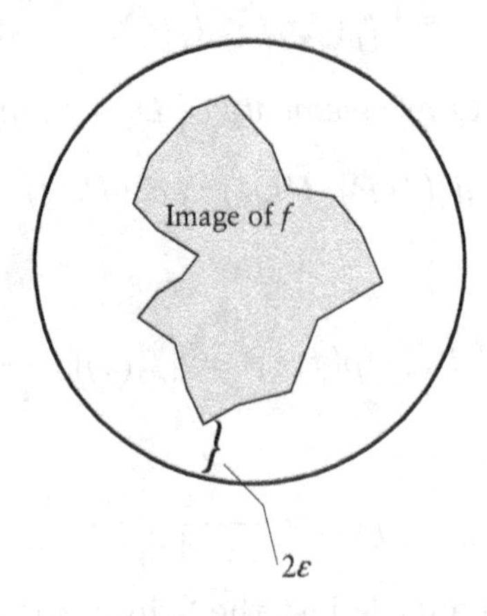

Figure 5.6: The Farkas/Ritt theorem.

By the preceding proposition, g is thus distance-decreasing in the Poincaré metric. Therefore

$$g^*\rho(z_0) \leq \rho(z_0).$$

Writing out the definition of g^* now yields

$$(1+\epsilon) \cdot f^*\rho(z_0) \leq \rho(z_0).$$

Note that this inequality holds for any $z_0 \in D$. But now if $\gamma : [a,b] \to D$ is any continuously differentiable curve, then we may conclude that

$$\ell_\rho(f_*\gamma) \leq (1+\epsilon)^{-1}\ell_\rho(\gamma).$$

If P and Q are elements of D and d is Poincaré distance, then we have that

$$d(f(P), f(Q)) \leq (1+\epsilon)^{-1} d(P,Q).$$

We see that f is a contraction in the Poincaré metric. Recall that in Proposition 5.19 we proved that the disc D is a complete metric space when equipped with the Poincaré metric. By the contraction mapping fixed-point theorem (see [LOS]), f has a unique fixed point. □

The next result shows how to find the fixed point, and, in effect, proves the contraction mapping fixed-point theorem in this special case.

COROLLARY 5.27 *If f is as in the theorem and P is the unique fixed point, then the iterates f, $f \circ f$, $f \circ f \circ f$, ... converge uniformly on compact sets to the constant function P.*

Proof: Let f^n denote the nth iterate of f. Let

$$\overline{\mathbf{B}}(P,R) = \{z \in D : d(z,P) \leq R\}$$

be the closed Poincaré metric ball with center P and Poincaré metric radius R. Then the theorem tells us that

$$f(\overline{\mathbf{B}}(P,R)) \subseteq \overline{\mathbf{B}}(P, R/(1+\epsilon))$$

and, more generally,

$$f^n(\overline{\mathbf{B}}(P,R)) \subseteq \overline{\mathbf{B}}(P, R/(1+\epsilon)^n). \tag{5.27.1}$$

Observe that

$$\bigcup_{j=1}^{\infty} \mathbf{B}(P,j) = D.$$

Now the exercise at the end of Section 2.1 tells us that these non-Euclidean balls are in fact open discs in the usual Euclidean topology.

Therefore every (Euclidean) compact subset K of D lies in some $\mathbf{B}(P, j)$. Then line (5.27.1) implies that

$$f^n(K) \subseteq \overline{\mathbf{B}}(P, j/(1+\epsilon)^n).$$

The result follows. □

5.3 The Schwarz Lemma According to Ahlfors

Capsule: Now we begin to develop the terminology and notation that will be needed for Ahlfors's version of the Schwarz lemma. In particular, we shall need a version of curvature that **(i)** is consistent with traditional geometric notions of curvature and **(ii)** fits naturally into the complex-variable context.

Definition 5.28 If $U \subseteq \mathbb{C}$ is a planar domain and ρ is a metric on U, then the *curvature* of the metric ρ at a point $z \in U$ is defined to be

$$\kappa_{U,\rho}(z) = \kappa(z) \equiv \frac{-\Delta \log \rho(z)}{\rho(z)^2}. \tag{5.28.1}$$

(Here zeros of $\rho(z)$ will result in singularities of the curvature function—κ is undefined at such points.)

Since ρ is twice continuously differentiable, this definition makes sense. It assigns a numerical quantity to each $z \in U$. The most important preliminary fact about κ is its conformal invariance:

PROPOSITION 5.29 *Let U_1 and U_2 be planar domains and $h : U_1 \to U_2$ a conformal map (in particular, h' never vanishes). If ρ is a metric on U_2 then*

$$\kappa_{U_1,h^*\rho}(z) = \kappa_{U_2,\rho}(h(z)), \qquad \forall z \in U_1.$$

Proof: We need to calculate:

$$\begin{aligned}\kappa_{U_1,h^*\rho}(z) &\equiv \frac{-\Delta \log[\rho(h(z)) \cdot \|h'(z)\|]}{[\rho(h(z)) \cdot |h'(z)|]^2} \\ &= \frac{-\Delta \log[\rho(h(z))] - \Delta[\log(\|h'(z)\|)]}{[\rho(h(z)) \cdot \|h'(z)\|]^2}.\end{aligned}$$

Now the second term in the numerator vanishes. We may further simplify the numerator to obtain that the last line equals

$$= \frac{-\left[\Delta \log \rho|_{h(z)}\right] \|h'(z)\|^2}{[\rho(h(z)) \cdot \|h'(z)\|]^2} = \frac{-\Delta \log \rho|_{h(z)}}{\rho(h(z))^2} = \kappa_{U_2,\rho}(h(z)). \quad \square$$

REMARK 5.30 In fact, the proof gives the slightly more general fact that if U_1, U_2 are domains and $f : U_1 \to U_2$ is a holomorphic map (not necessarily one-to-one or onto), then the following holds. If ρ is a metric on U_2, then

$$\kappa_{f^*\rho}(z) = \kappa_\rho(f(z))$$

at every point $z \in U_1$ for which $f'(z) \neq 0$ and $\rho(f(z)) \neq 0$. Usually this is all points except for a discrete set.

From the point of view of geometry, any differential quantity that is invariant is automatically of great interest. But why do we call κ "curvature"? The answer is that a standard construction in Riemannian geometry assigns a quantity called Gaussian curvature to any Riemannian metric. In the setting of classical Euclidean geometry this quantity coincides with our intuitive notion of what curvature ought to be. In more abstract settings, the structural equations of Cartan lead to a quantity called curvature, which is invariant but has considerably less intuitive content. The appendix at the end of this section gives a complete derivation—in the complex plane—of the above notion κ of curvature from more classical considerations of Gaussian curvature. One of the pleasant facts about the use of elementary differential geometry in complex function theory is that one does not require all the geometric machinery that leads to the formula for κ in order to make use of κ.

Let us begin our study of curvature by calculating the curvature of the Euclidean metric.

Example 5.31 Let U be a planar domain equipped with the Euclidean metric $\rho(z) \equiv 1$. It follows from the definition that $\kappa(z) \equiv 0$. This should be expected, for the Euclidean metric does not change from point to point.

Example 5.32 The metric

$$\sigma(z) = \frac{2}{1 + \|z\|^2}$$

on $\mathbb{C}$ is called the *spherical metric*. This is in fact the metric that the plane inherits from the sphere in Euclidean three space by way of the stereographic projection. A straightforward calculation shows that the curvature of σ is identically 1. It is a basic theorem of Riemannian geometry that the sphere is the only manifold of constant positive curvature. In fact, the study of manifolds of constant curvature comprises an entire subject in itself (see [WOL]).

Next we calculate the curvature of the Poincaré metric on the disc, which calculation will be of great utility for us. Notice that, since any point of the disc may be moved to any other by a Möbius transformation, and since curvature is a conformal invariant, we will expect the curvature function to be constant.

PROPOSITION 5.33 *Consider the disc D equipped with the Poincaré metric. For any point $z \in D$ it holds that $\kappa(z) = -4$.*

Proof: We notice that

$$-\Delta \log \rho(z) = \Delta \log(1 - \|z\|^2).$$

Now we write $\Delta = 4(\partial/\partial z)(\partial/\partial \bar{z})$ and $\|z\|^2 = z \cdot \bar{z}$ to obtain that this last expression equals

$$-\frac{4}{(1 - \|z\|^2)^2}.$$

It follows that $\kappa(z) = -4$. □

The fact that the Poincaré metric on the disc has constant negative curvature turns out to be a special property of the disc. Now we return to the Schwarz lemma. The constancy of curvature follows immediately from the transitivity of the automorphism group. The negativity is a special feature of the geometry.

Ahlfors [AHL1] first realized that the Schwarz lemma is really an inequality about curvature. In his annotations to his collected works he modestly asserts that, "This is an almost trivial fact and anybody who sees the need could prove it at once." However, he goes on to say (most correctly) that this point of view has been a decisive influence in modern function theory. It is a good place to begin our understanding of curvature. Here is Ahlfors's version of Schwarz's lemma:

Prelude: This is a theorem that certainly bears some thought. You should consider some example domains U, such as a larger disc and an annulus. Decide on a metric σ and see what the theorem says.

THEOREM 5.34 *Let the disc $D = D(0,1)$ be equipped with the Poincaré metric ρ, and let U be a planar domain equipped with a metric σ. Assume that, at all points of U, σ has curvature not exceeding -4. If $f : D \to U$ is holomorphic, then we have*

$$f^*\sigma(z) \leq \rho(z), \qquad \forall z \in D.$$

Proof: We follow the argument in [MIS]. Let $0 < r < 1$. On the disc $D(0,r)$ define the metric

$$\rho_r(z) = \frac{r}{r^2 - z^2}.$$

Then straightforward calculations (or change of variable) show that ρ_r is the analogue of the Poincaré metric for $D(0,r)$: it has constant curvature -4 and is invariant under conformal maps. Define

$$v = \frac{f^*\sigma}{\rho_r}.$$

Observe that v is continuous and nonnegative on $D(0,r)$ and that $v \to 0$ when $|z| \to r$ (since $f^*\sigma$ is bounded above on $\bar{D}(0,r) \subset\subset D$ while $\rho_r \to \infty$). It follows that v attains a maximum value M at some point $\tau \in D(0,r)$. We will show that $M \leq 1$; hence that $v \leq 1$ on $D(0,r)$. Letting $r \to 1^-$ then finishes the proof.

If $f^*\sigma(\tau) = 0$, then $v \equiv 0$. So we may suppose that $f^*\sigma(\tau) > 0$. Therefore $\kappa_{f^*\sigma}$ is defined at τ. By hypothesis,

$$\kappa_{f_*\sigma} \leq -4.$$

Since $\log v$ has a maximum at τ, we have

$$\begin{aligned} 0 &\geq \Delta \log v(\tau) \\ &= \Delta \log f^*\sigma(\tau) - \Delta \log \rho_r(\tau) \\ &= -\kappa_{f_*\sigma}(\tau) \cdot (f^*\sigma(\tau))^2 + \kappa_{\rho_r}(\tau) \cdot (\rho_r(\tau))^2 \\ &\geq 4(f^*\sigma(\tau))^2 - 4(\rho_r(\tau))^2. \end{aligned}$$

This gives

$$\frac{f^*\sigma(\tau)}{\rho_r(\tau)} \leq 1,$$

or

$$M \leq 1.$$

This is the desired estimate. □

Observe that the usual Schwarz lemma is a corollary of Ahlfors's version: we take U to be the disc with σ being the Poincaré metric. Let f be a holomorphic mapping of D to $U = D$ such that $f(0) = 0$. Then σ satisfies the hypotheses of the theorem and the conclusion is that

$$f^*\sigma(0) \leq \rho(0).$$

Unraveling the definition of $f^*\sigma$ yields

$$\|f'(0)\| \cdot \sigma(f(0)) \leq \rho(0).$$

But $\sigma = \rho$ and $f(0) = 0$, so this last inequality becomes

$$\|f'(0)\| \leq 1.$$

Exercise: The property

$$d(f(P), f(Q)) \leq d(P, Q),$$

where d is the Poincaré distance, is called the *distance-decreasing property* of the Poincaré metric. Use this property to give a geometric proof of the other inequality in the classical Schwarz lemma. □

With a little more notation, we can obtain a more powerful version of the Ahlfors/Schwarz lemma. Let $D(0, \alpha)$ be the open disc of radius α and center 0. For $A > 0$, define the metric ρ_α^A on $D(0, \alpha)$ by

$$\rho_\alpha^A(z) = \frac{2\alpha}{\sqrt{A}(\alpha^2 - \|z\|^2)}.$$

This metric has constant curvature $-A$. Now we obtain the following more general version of Ahlfors's Schwarz lemma.

Prelude: This generalized version of Ahlfors's Schwarz lemma allows for more general domains. Its proof is not much different, but it is a decisively more useful result.

THEOREM 5.35 *Let U be a planar domain that is equipped with a metric σ whose curvature is bounded above by a negative constant $-B$. Then every holomorphic function $f : D(0, \alpha) \to U$ satisfies*

$$f^*\sigma(z) \leq \frac{\sqrt{A}}{\sqrt{B}} \rho_\alpha^A(z), \qquad \forall z \in D(0, \alpha).$$

It is a good exercise to construct a proof of this more general result, modeled on the proof of Theorem 5.34.

In the next section we shall see several elegant applications of Theorems 5.34 and 5.35.

APPENDIX: A Curvature Calculation

Here we give a brief presentation of the connection between the calculus notion of curvature (see [THO]) and the more abstract notion of curvature which leads to the definition of κ in Chapter 2. A more detailed treatment of some of these matters may be found in [ONE].

First, a word about notation. We use the language of differential forms consistently in this appendix. On the one hand, classically trained analysts are often uncomfortable with this language. On the other hand, the best way to learn the language is to use it. And the context of curvature calculations on plane domains may in fact be the simplest nontrivial context in which differential forms can be profitably used. In any event, this appendix would be terribly clumsy if we did not use forms, so the decision is essentially automatic. All necessary background on differential forms may be found in [RU1] or in [ONE].

An Intrinsic Look at Curvature

First we recall the concept of curvature for a smooth, two-dimensional surface $M \subseteq \mathbf{R}^3$. All of our calculations are local, so it is convenient to think of M as parametrized by two coordinate functions over a connected open set $U \subseteq \mathbf{R}^2$:

$$U \ni (u,v) \xrightarrow{p} (x_1(u,v), x_2(u,v), x_3(u,v)) \in M.$$

We require that the matrix

$$\begin{pmatrix} \frac{\partial x_1}{\partial u} & \frac{\partial x_2}{\partial u} & \frac{\partial x_3}{\partial u} \\ \frac{\partial x_1}{\partial v} & \frac{\partial x_2}{\partial v} & \frac{\partial x_3}{\partial v} \end{pmatrix}$$

have rank 2 at each point of U. The vectors given by the rows of this matrix span the tangent plane to M at each point. Applying the Gram-Schmidt orthonormalization procedure to these row vectors, and shrinking U, M if necessary, we may create vector fields

$$E_1 : M \longrightarrow \mathbb{R}^3$$
$$E_2 : M \longrightarrow \mathbb{R}^3$$

such that $E_1(x_1, x_2, x_3)$ and $E_2(x_1, x_2, x_3)$ are orthonormal and tangent to M at each $P = (x_1, x_2, x_3) \in M$. Denote by $T_P(M)$ the collection of linear combinations

$$aE_1(x_1, x_2, x_3) + bE_2(x_1, x_2, x_3), \qquad a, b \in \mathbf{R}.$$

We call $T_P(M)$ the *tangent space* to M at P.

Let $E_3(P)$ be the unit normal to M at P given by the cross product $E_1(P) \times E_2(P)$. The functions E_1, E_2, E_3 are smooth *vector fields* on M; they assign to each $P \in M$ a triple of orthonormal vectors.

Let $\delta_1, \delta_2, \delta_3$ denote the standard basis for vectors in $\mathbf{R}^3$:

$$\begin{aligned} \delta_1 &= (1, 0, 0), \\ \delta_2 &= (0, 1, 0), \\ \delta_3 &= (0, 0, 1). \end{aligned}$$

(Many calculus books call these vectors **i**, **j**, and **k**.) Then we may write

$$E_i = \sum_j a_{i,j}(x_1, x_2, x_3)\delta_j, \qquad i = 1, 2, 3.$$

The matrix

$$\mathcal{A} \equiv (a_{i,j})_{i,j=1}^3,$$

where the $a_{i,j}$ are *functions* of the space variables, is called the *attitude matrix* of the frame E_1, E_2, E_3. Since $\mathcal{A}$ transforms one orthonormal frame to another, $\mathcal{A}$ is an orthogonal matrix. Hence

$$\mathcal{A}^{-1} = {}^t\mathcal{A}.$$

Definition 5.36 If $P \in M, v \in T_P(M)$, and f is a smooth function on M, then define

$$D_v f(P) = \frac{d}{dt} f \circ \varphi(t)\bigg|_{t=0},$$

where φ is any smooth curve in M such that $\varphi(0) = P$ and $\varphi'(0) = v$. One checks that this definition is independent of the choice of φ.

Definition 5.37 If

$$\alpha : M \longrightarrow \mathbf{R}^3$$

is a vector field on M,

$$\alpha(P) = \alpha_1(P)\delta_1 + \alpha_2(P)\delta_2 + \alpha_3(P)\delta_3,$$

and $v \in T_P(M)$, then define

$$\nabla_v \alpha(P) = (D_v\alpha_1)(P)\delta_1 + (D_v\alpha_2)(P)\delta_2 + (D_v\alpha_3)(P)\delta_3.$$

The operation ∇_v is called *covariant differentiation* of the vector field α.

Definition 5.38 If $P \in M$, $v \in T_P(M)$, we define the *shape operator* (or *Weingarten map*) for M at P to be

$$S_P(v) = -\nabla_v E_3(P).$$

LEMMA 5.39 *We have that $S_P(v) \in T_P(M)$.*

Proof: Now $E_3 \cdot E_3 \equiv 1$ hence

$$\begin{aligned} 0 &= D_v(E_3 \cdot E_3)\Big|_P \\ &= (2\nabla_v E_3) \cdot E_3\Big|_P \\ &= -2S_P(v) \cdot E_3(P). \end{aligned}$$

Hence $S_P(v) \perp E_3(P)$, so $S_P(v) \in T_P(M)$. □

Notice that the shape operator assigns to each $P \in M$ a linear operator S_P on the two-dimensional tangent space $T_P(M)$.

We can express this linear operator as a matrix with respect to the basis $E_1(P)$, $E_2(P)$. So S assigns to each $P \in M$ a 2×2 matrix $\mathcal{M}_P$.

The linear operator S_P measures the rate of change of E_3 in any tangent direction v. It can be shown that $\mathcal{M}_P$ is diagonalizable. The (real) eigenvectors of $\mathcal{M}_P$ correspond to the *principal curvatures* of M at P (these are the directions of greatest and least curvature), and the corresponding eigenvalues measure the amount of curvature in those directions.

The preceding observations about $\mathcal{M}_P$ motivate the following definition.

Definition 5.40 The *Gaussian curvature* $\kappa(P)$ of M at a point $P \in M$ is the determinant of $\mathcal{M}_P$—the product of the two eigenvalues.

Our aim is to express $\kappa(P)$ in terms of the intrinsic geometry of M, without reference to E_3—that is, without reference to the way that M is situated in space.

To this end, we define *covector fields* θ_i which are dual to the vector fields E_i :

$$\theta_i E_j(P) = \delta_{ij}.$$

Then θ_i may be expressed as linear combinations of the standard basis covectors dx_1, dx_2, dx_3. Indeed, if $\mathcal{A} = (a_{i,j})$ is the attitude matrix then

$$\theta_i = \Sigma a_{i,j} dx_j$$

(remember that $\mathcal{A}^{-1} = {}^t\mathcal{A}$). Thus, for each i, θ_i is a differential form; and the standard calculus of differential forms—including exterior differentiation—applies.

From now on, we restrict attention to one- and two-forms acting on tangent vectors to M. Thus any one-form α may be expressed as

$$\alpha = \alpha(E_1)\theta_1 + \alpha(E_2)\theta_2$$

and any two-form β may be expressed as

$$\beta = \beta(E_1, E_2)\theta_1 \wedge \theta_2.$$

Now we define covector fields $U_{i,j}$, $i, j \in \{1, 2, 3\}$, by the formula

$$U_{i,j}(v) = (\nabla_v E_i) \cdot E_j(P).$$

Here "$\cdot$" is the Euclidean dot product. We think of $U_{i,j}$ as a differential 1-form. Notice that, since $E_i \cdot E_j \equiv \delta_{i,j}$, we have, for $v \in T_P(M)$, that

$$\begin{aligned} 0 &= D_v(E_i \cdot E_j) \\ &= (\nabla_v E_i) \cdot E_j + E_i \cdot (\nabla_v E_j) \\ &= U_{i,j}(v) + U_{j,i}(v). \end{aligned}$$

Thus

$$U_{i,j} = -U_{j,i}.$$

In particular,

$$U_{i,i} = 0.$$

If $v \in T_P(M)$, then it is easy to check, just using linear algebra, that

$$\nabla_v E_i = \sum_j U_{i,j}(v) E_j, \qquad 1 \le i \le 3.$$

We call the $U_{i,j}$ the *connection forms* for M. We can now express the shape operator in terms of these connection forms.

PROPOSITION 5.41 *Let $P \in M$ and $v \in T_P(M)$. Then*

$$S_P(v) = U_{1,3}(v)E_1(P) + U_{2,3}(v)E_2(P).$$

Proof: We have that

$$\begin{aligned}S_P(v) &= -\nabla_v E_3 \\ &= -\sum_{j=1}^{3}(\nabla_v E_3 \cdot E_j)E_j \\ &= -\sum_{j=1}^{3} U_{3,j}(v)E_j \\ &= U_{1,3}(v)E_1 + U_{2,3}(v)E_2,\end{aligned}$$

since $U_{3,3} = 0$. □

Now we can express Gaussian curvature in terms of the $U_{i,j}$.

PROPOSITION 5.42 *We have that*

$$U_{1,3} \wedge U_{2,3} = \kappa\ \theta_1 \wedge \theta_2.$$

Proof: We need to calculate $\mathcal{M}_P$ in terms of the $U_{i,j}$. Proposition 5.41 gives that

$$S_P(E_1) = U_{1,3}(E_1)E_1 + U_{2,3}(E_1)E_2$$

and

$$S_P(E_2) = U_{1,3}(E_2)E_1 + U_{2,3}(E_2)E_2$$

so the matrix of S_P, in terms of the basis E_1, E_2, is

$$\mathcal{M}_P = \begin{pmatrix} U_{1,3}(E_1) & U_{2,3}(E_1) \\ U_{1,3}(E_2) & U_{2,3}(E_2) \end{pmatrix}.$$

We know that $U_{1,3} \wedge U_{2,3}$, being a two-form, can be written as $\lambda \cdot \theta_1 \wedge \theta_2$. On the other hand,

$$\begin{aligned}\kappa &= \det \mathcal{M}_P \\ &= U_{1,3}(E_1)U_{2,3}(E_2) - U_{1,3}(E_2)U_{2,3}(E_1) \\ &= (U_{1,3} \wedge U_{2,3})(E_1, E_2) \\ &= \lambda.\end{aligned}$$

Therefore

$$U_{1,3} \wedge U_{2,3} = \lambda\ \theta_1 \wedge \theta_2 = \kappa\ \theta_1 \wedge \theta_2.$$ □

Our goal now is to express κ using only those $U_{i,j}$ with $i \neq 3,\ j \neq 3$. For this we require a technical lemma about the attitude matrix.

LEMMA 5.43 *We have that*

$$U_{i,j} = \sum_k a_{j,k} da_{i,k}, \qquad 1 \le i,\ j \le 3.$$

Proof: If $v \in T_P(M)$ then

$$U_{i,j}(v) = \nabla_v E_i \cdot E_j(P).$$

But

$$E_i = \sum_k a_{i,k}\delta_k.$$

Therefore

$$\nabla_v E_i = \sum_k (D_v a_{i,k})\delta_k.$$

Then

$$\begin{aligned} U_{i,j}(v) &\equiv \nabla_v E_i \cdot E_j \\ &= \left(\sum_k (D_v a_{i,k})\delta_k\right) \cdot \left(\sum_k a_{j,k}\delta_k\right) \\ &= \sum_k (D_v a_{i,k}) a_{j,k} \\ &= \sum_k da_{i,k}(v) a_{j,k}. \end{aligned}$$

As a result,

$$U_{i,j} = \sum_k a_{j,k} da_{i,k}. \qquad \square$$

Now we have reached a milestone. We can derive the important *Cartan structural equations*, which are the key to our intrinsic formulas for curvature.

Prelude: The Cartan structural equations are a very basic part of curvature theory in elementary differential geometry. You will see below the central role that they play in our reasoning.

THEOREM 5.44 *We have*

$$d\theta_i = \sum_j U_{i,j} \wedge \theta_j, \tag{5.44.1}$$

and

$$dU_{i,j} = \sum_k U_{i,k} \wedge U_{k,j}. \tag{5.44.2}$$

Proof: We have that

$$\theta_i = \sum_j a_{i,j} dx_j$$

hence

$$d\theta_i = \Sigma da_{i,j} \wedge dx_j.$$

Since the attitude matrix $\mathcal{A}$ is orthogonal, we may solve the equations in Lemma 5.43 for $da_{i,k}$ in terms of $U_{i,j}$. Thus

$$da_{i,j} = \Sigma U_{i,k} a_{k,j}.$$

Substituting this identity into our last formula gives

$$\begin{aligned} d\theta_i &= \sum_j \left[\left(\sum_k U_{i,k} a_{k,j} \right) \wedge dx_j \right] \\ &= \sum_k \left[U_{i,k} \wedge \sum_j a_{kj} dx_j \right] \\ &= \sum_k U_{i,k} \wedge \theta_k. \end{aligned}$$

This is the first structural equation.

For the second equation, notice that the formula

$$U_{i,j} = \Sigma da_{i,k} a_{j,k}$$

implies

$$dU_{i,j} = -\Sigma da_{i,k} \wedge da_{j,k}.$$

On the other hand,

$$\begin{aligned} \sum_k U_{i,k} \wedge U_{k,j} &= \sum_k \left(\sum_\ell da_{i,\ell} a_{k,\ell} \right) \wedge \left(\sum_m da_{k,m} a_{j,m} \right) \\ &= \sum_k \left(\sum_\ell da_{i,\ell} a_{k,\ell} \right) \wedge \left(- \sum_m da_{j,m} a_{k,m} \right) \\ &= -\left(\sum_k a_{k,\ell} a_{k,m} \right) \cdot \left(\sum_{\ell,m} da_{i,\ell} \wedge da_{j,m} \right) \\ &= -\sum_m da_{i,m} \wedge da_{j,m}. \end{aligned}$$

In the penultimate equality we have used the fact that $\mathcal{A}^{-1} = {}^t\mathcal{A}$. The result now follows. □

The following corollary will prove critical.

COROLLARY 5.45 *We have*

$$dU_{1,2} = -\kappa\, \theta_1 \wedge \theta_2.$$

Proof: The second structural equation gives

$$\begin{aligned} dU_{1,2} &= \Sigma U_{1,k} \wedge U_{k,2} \\ &= U_{1,1} \wedge U_{1,2} + U_{1,2} \wedge U_{2,2} + U_{1,3} \wedge U_{3,2}. \end{aligned}$$

Only the third summand doesn't vanish. But we calculated in Proposition 5.42 that it equals $-\kappa\, \theta_1 \wedge \theta_2$. □

The corollary has been our main goal in this subsection. It gives an intrinsic way to calculate Gaussian curvature in the classical setting; hence it gives a way to *define* Gaussian curvature in more abstract settings. We now proceed to develop this more abstract point of view.

Curvature on Planar Domains

Let $U \subseteq \mathbf{C}$ be a domain which is equipped with a metric ρ. Assume, for simplicity, that $\rho(z) > 0$ at all points of U. Define

$$E_1 \equiv \frac{(1,0)}{\rho} \qquad \text{and} \qquad E_2 \equiv \frac{(0,1)}{\rho}.$$

Then

$$\theta_1 = \rho\, dx \qquad \text{and} \qquad \theta_2 = \rho\, dy$$

are the dual covector fields. We define $U_{i,j}$ according to the first structure equation:

$$\begin{aligned} d\theta_1 &= U_{1,2} \wedge \theta_2, \\ d\theta_2 &= U_{2,1} \wedge \theta_1. \end{aligned}$$

We define Gaussian curvature according to the Corollary to Theorem 5.44:

$$dU_{1,2} = -\kappa\, \theta_1 \wedge \theta_2.$$

One can check that these definitions are independent of the choice of frame E_1, E_2, but this is irrelevant for our purposes.

We conclude this appendix by proving that the definition of curvature which we just elicited from the structural equations coincides with the one given in Section 5.4. First, by the way that we've defined θ_1 and θ_2, we have

$$\begin{aligned} d\theta_1 &= d\rho \wedge dx \\ &= (\rho_x dx + \rho_y dy) \wedge dx \\ &= \rho_y dy \wedge dx \\ &= -\frac{\rho_y}{\rho} dx \wedge \rho\, dy \\ &= -\frac{\rho_y}{\rho} dx \wedge \theta_2. \end{aligned}$$

Similarly,

$$\begin{aligned} d\theta_2 &= d\rho \wedge dy \\ &= (\rho_x dx + \rho_y dy) \wedge dy \\ &= \rho_x dx \wedge dy \\ &= -\frac{\rho_x}{\rho} dy \wedge \rho\, dx \\ &= -\frac{\rho_x}{\rho} dy \wedge \theta_1. \end{aligned}$$

Comparison with the first structural equation gives

$$U_{1,2} = -\frac{\rho_y}{\rho} dx + \tau\, dy$$

and

$$U_{1,2} = -U_{2,1} = -\left(-\frac{\rho_x}{\rho} dy\right) + \sigma\, dx,$$

for some unknown functions τ and σ.

The only way these equations can be consistent is if

$$U_{1,2} = -\frac{\rho_y}{\rho} dx + \frac{\rho_x}{\rho} dy.$$

Thus

$$\begin{aligned} dU_{1,2} &= -\frac{\partial}{\partial y}\left(\frac{\rho_y}{\rho}\right) dy \wedge dx + \frac{\partial}{\partial x}\left(\frac{\rho_x}{\rho}\right) dx \wedge dy \\ &= \left(-\frac{\rho_{yy}}{\rho} + \frac{\rho_y \rho_y}{\rho^2}\right) dy \wedge dx + \left(\frac{\rho_{xx}}{\rho} - \frac{\rho_x \rho_x}{\rho^2}\right) dx \wedge dy \\ &= \frac{1}{\rho^2}(\rho \Delta \rho - (\rho_y)^2 - (\rho_x)^2) dx \wedge dy \\ &= \frac{1}{\rho^4}(\rho \Delta \rho - (\rho_y)^2 - (\rho_x)^2) \theta_1 \wedge \theta_2. \end{aligned}$$

The second structural equation now implies that

$$\kappa = -\frac{1}{\rho^4}(\rho\Delta\rho - |\nabla\rho|^2).$$

On the other hand, in Section 5.4 we defined

$$\kappa = -\frac{\Delta \log \rho}{\rho^2}.$$

We have

$$\frac{\partial}{\partial \bar{z}} \log \rho = \frac{1}{\rho}\frac{\partial \rho}{\partial \bar{z}}$$

and

$$\begin{aligned}
\Delta \log \rho &= 4\frac{\partial^2}{\partial z \partial \bar{z}} \log \rho \\
&= 4\left(-\frac{1}{\rho^2}\frac{\partial \rho}{\partial z}\frac{\partial \rho}{\partial \bar{z}} + \frac{1}{\rho}\frac{\partial^2 \rho}{\partial z \partial \bar{z}}\right) \\
&= -\frac{1}{\rho^2}\cdot|\nabla\rho|^2 + \frac{1}{\rho}\cdot\Delta\rho.
\end{aligned}$$

It follows that, according to the definition in Section 5.4,

$$\kappa = -\frac{1}{\rho^4}(\rho\Delta\rho - |\nabla\rho|^2).$$

Thus, as claimed, the definition of curvature that we studied earlier is equal to that which arises from the structural equations.

5.4 Another Look at Schwarz's Lemma

Capsule: In fact there are many ways to think about the Schwarz lemma phenomenon. Some are in the vein of the Schwarz-Pick lemma, and others are definitely new and different.

We explore some of these in the present section.

Here we present some fascinating, but less well known, versions of the Schwarz lemma concept.

PROPOSITION 5.46 *Let f be holomorphic on $D(0, r)$, and assume that $|f(z)| \le M$ for all z. Then*

$$\left|\frac{f(z) - f(w)}{z - w}\right| \le \frac{2Mr}{|r^2 - \bar{z}w|}.$$

Proof: Define

$$g(z) = \frac{f(rz)}{M}.$$

Then $g : D \to D$, and we may apply Schwarz-Pick to g. The result is

$$\left| \frac{g(z) - g(w)}{1 - \overline{g(z)} g(w)} \right| \leq \left| \frac{z - w}{1 - \overline{z} w} \right|,$$

which translates to

$$\left| \frac{f(rz)/M - f(rw)/M}{1 - \overline{f(rz)/M} \cdot f(rw)/M} \right| \leq \left| \frac{z - w}{w - \overline{z} w} \right|.$$

Doing some elementary algebra and replacing $z \mapsto z/r$, $w \mapsto w/r$ yields

$$\left| \frac{f(z) - f(w)}{1 - [\overline{f(z)} f(w)]/M^2} \right| \leq M \cdot \left| \frac{r(z - w)}{r^2 - \overline{z} w} \right|.$$

This may be rearranged as

$$\left| \frac{f(z) - f(w)}{z - w} \right| \leq \frac{rM \left| 1 - (\overline{f(z)} f(w))/M^2 \right|}{|r^2 - \overline{z} w|}$$

and the right-hand side here can obviously be majorized by

$$\frac{2Mr}{|r^2 - \overline{z} w|}.$$

That completes the argument. □

The next result, due to Lindelöf, gives a means for majorizing $|f(z)|$, in the classical Schwarz lemma setting, *without* assuming that $f(0) = 0$. We can think of it as another way to view the Schwarz-Pick phenomenon.

PROPOSITION 5.47 *Let $f : D \to D$ be holomorphic. Then*

$$|f(z)| \leq \frac{|f(0)| + |z|}{1 + |f(0)||z|}.$$

Proof: We first note that, for complex numbers $a, b \in D$,

$$\begin{aligned}\left|\frac{a-b}{1-\overline{a}b}\right|^2 &= 1 - \frac{(1-|a|^2)(1-|b|^2)}{|1-\overline{a}b|^2} \\ &\geq 1 - \frac{(1-|a|^2)(1-|b|^2)}{(1-|a||b|)^2} \\ &= \frac{(|a|-|b|)^2}{(1-|a||b|)^2}.\end{aligned}$$

Now we use this inequality to calculate, for $z \in D$, that

$$\frac{|f(z)| - |f(0)|}{1 - |f(0)||f(z)|} \leq \frac{|f(z) - f(0)|}{|1 - \overline{f(0)}f(z)|}.$$

By Schwarz-Pick this last righthand quantity is not greater than

$$\frac{|z-0|}{|1-\overline{0}z|} = |z|.$$

Thus

$$|f(z)| - |f(0)| \leq |z| - |z| \cdot |f(0)| \cdot |f(z)|.$$

Rearranging the inequality gives

$$|f(z)| \leq \frac{|f(0)| + |z|}{1 + |z||f(0)|},$$

which is the desired result. □

Chapter 6

Harmonic Measure

Prologue: It is unfortunate that most graduate complex analysis courses do not treat harmonic measure. Indeed, most complex analysis texts do not cover the topic.

The term "harmonic measure" was introduced by R. Nevanlinna in the 1920s. But the ideas were anticipated in work of the Riesz brothers and others. It is still studied intensively today.

Harmonic measure is a natural outgrowth of the study of the Dirichlet problem. It is a decisive tool for measuring the growth of functions, and to seeing how they bend. It has been used in such diverse applications as the corona problem and in mapping problems. Harmonic measure has particularly interesting applications in probability theory—especially in the consideration of diffustions and Brownian motion.

It requires a little sophistication to appreciate harmonic measure. And calculating the measure—working through the examples—requires some elbow grease. But the effort is worth it and the results are substantial.

This chapter gives a basic introduction to the idea of harmonic measure.

6.1 The Idea of Harmonic Measure

Capsule: Harmonic measure is a way of manipulating the data that comes from the Dirichlet problem. It is a way of measuring the geometry of a domain, and can be used to control the growth of harmonic functions.

Let $\Omega \subseteq \mathbb{C}$ be a domain with boundary consisting of finitely many Jordan curves (we call such a domain a *Jordan domain*). Let E be a finite union of arcs in $\partial\Omega$ (we allow the possibility of an entire connected boundary component—a simple, closed curve—to be one of these arcs). Then the *harmonic measure* of E at the point $z \in \Omega$ with respect to Ω is the value at z of the bounded harmonic function ω on Ω with boundary limit 1 at points of E and boundary limit 0 at points of $\partial\Omega \setminus E$ (except possibly at the endpoints of the arcs that make up E). We denote the harmonic measure by $\omega(z, \Omega, E)$.

The first question to ask about harmonic measure is that of existence and uniqueness: does harmonic measure always *exist*? If it does exist, is it unique? The answer to both these queries is "yes." Let us see why.

Fix a Jordan domain Ω. It is a classical result (see [AHL2], as well as our Section 1.5) that there is a conformal mapping $\Phi : \Omega \to U$ of Ω to a domain U bounded by finitely many circles. If $F \subseteq \partial\Omega$ consists of finitely many circular arcs, then it is obvious that the harmonic measure of F exists. It would just be the Poisson integral of the characteristic function of F, and that Poisson integral can be shown to exist by elementary conformal mapping arguments (i.e., mapping to the disc). Now a classic result of Carathéodory guarantees that Φ and its inverse extend continuously to the respective boundaries, so that the extended function is a homeomorphism of the closures. Thus we may pull the harmonic measure back from U to Ω via Φ.

As for uniqueness: if Ω is bounded, then uniqueness follows from the standard maximum principle. For the general case, we need an extended maximum principle that is due to Lindelöf:

PROPOSITION 6.1 *Let Ω be a domain whose boundary is not a finite set. Let u be a real-valued harmonic function on Ω, and assume that there is a real constant $M > 0$ such that*

$$u(z) \leq M, \qquad \text{for } z \in \Omega.$$

Suppose that m is a real constant and that

$$\limsup_{z \to \zeta} u(z) \leq m \tag{6.1.1}$$

for all except possibly finitely many points $\zeta \in \partial\Omega$. Then $u(z) \leq m$ for all $z \in \Omega$.

REMARK 6.2 Observe that any Jordan domain certainly satisfies the hypotheses of the proposition. Furthermore, by classical results coming from potential theory (see [AHL2]), any domain on which the Dirichlet problem is solvable will satisfy the hypotheses of the theorem. Note, in particular, that Ω certainly need not be bounded.

We conclude by noting that this result is a variant of the famous Phragmen-Lindelöf theorem (see [RUD2]).

Proof of Proposition 6.1: First assume that Ω is bounded. We will remove this extra hypothesis later. Let the diameter of Ω be d. Let the exceptional boundary points (at which (6.1.1) does not hold) be called $\zeta_1, \ldots, \zeta_k$. Let $\epsilon > 0$. We then may apply the ordinary textbook maximum principle (see [GRK1]) to the auxiliary function

$$h(z) = u(z) + \epsilon \sum_{j=1}^{k} \log \frac{|z - \zeta_j|}{d}.$$

Notice that h (instead of u) satisfies the hypotheses of the proposition at *every boundary point*. So $h \leq m$ on all of Ω. Then we let $\epsilon \to 0$ and the result follows.

Now consider the case of Ω unbounded. If Ω has an exterior point (i.e., a point in the interior of the complement of the closure of Ω), then we may apply an inversion and reduce to the case in the preceding paragraph.

Finally, if Ω has no exterior point, then let $R > |\zeta_j|$ for all $j = 1, \ldots, k$. Let

$$\Omega_1 = \{z \in \Omega : |z| < R\}$$

and

$$\Omega_2 = \{z \in \Omega : |z| > R\}.$$

These may not be domains (i.e., they could be disconnected), but they are certainly open sets. Let

$$S = \{z \in \Omega : |z| = R\}.$$

If $u \leq m$ on S, then we apply the result of the first paragraph on Ω_1 and the result of the second paragraph on Ω_2.

If it is *not* the case that $u \leq m$ on S, then u will have a maximum N on S with $N > m$. But this will be a maximum for u on all of Ω. Since $u \leq m$ at the ends of the arcs of S, it follows that u actually achieves the maximum N on S. Therefore, by the usual maximum principle, u is identically equal to the constant N. But then the boundary condition (6.1.1) cannot obtain. That is a contradiction. □

It is important to realize that, if Ω has reasonably nice boundary, then $\omega(z, \Omega, E)$ is nothing other than the Poisson integral of the characteristic function of E. (We used this fact in our discussion of the existence of harmonic measure.) In any event, by the maximum principle the function ω takes real values between 0 and 1. Therefore $0 < \omega(z, \Omega, E) < 1$.

One of the reasons that harmonic measure is important is that it is a conformal invariant. Essential to this fact is Carathéodory's theorem, that a conformal map of Jordan domains will extend continuously to the closures. We now formulate the invariance idea precisely.

PROPOSITION 6.3 *Let Ω_1, Ω_2 be domains with boundaries consisting of Jordan curves in $\mathbb{C}$ and $\varphi : \Omega_1 \to \Omega_2$ a conformal map. If $E_1 \subseteq \partial\Omega_1$ is an arc and $z \in \Omega_1$ then*

$$\omega(z, \Omega_1, E_1) = \omega(\varphi(z), \Omega_2, \varphi(E_1)).$$

Here $\varphi(E_1)$ is well defined by Carathéodory's theorem.

Proof: Let h denote the harmonic function $\omega(\varphi(z), \Omega_2, \varphi(E_1))$. Then $h \circ \varphi$ equals harmonic measure for Ω_1 (by Carathéodory's theorem). □

REMARK 6.4 It is worth recording here an important result of F. and M. Riesz (see [KOO, page 72]). Let $\Omega \subseteq \mathbb{C}$ be a Jordan domain bounded by a rectifiable boundary curve. Let $\varphi : D \to \Omega$ be a conformal mapping. If $E \subseteq \partial D$ has Lebesgue linear measure zero, then $\varphi(E)$ also has Lebesgue linear measure zero.

6.2 Some Examples

Capsule: An idea like harmonic measure is a bit hollow without some concrete examples. In this section, we produce several very concrete instances.

Harmonic measure helps us to understand the growth and value distribution of harmonic and holomorphic functions on Ω. It has become a powerful analytic tool. We begin to understand harmonic measure by first calculating some examples.

Example 6.5 Let $\mathcal{U}$ be the upper-half-plane. Let E be an interval $[-T, T]$ on the real axis, centered at 0. Let us calculate $\omega(z, \mathcal{U}, E)$ for $z = x + iy \cong (x, y) \in \Omega$.

It is a standard fact (see [GRK1]) that the Poisson kernel for $\mathcal{U}$ is

$$P(x, y) = \frac{1}{\pi} \cdot \frac{y}{x^2 + y^2}.$$

Here we use the traditional real notation for the kernel. In other words, the harmonic function Ω that we seek is given by

$$\begin{aligned} \omega(x + iy) &= \frac{1}{\pi} \int_{-\infty}^{\infty} \chi_{[-T,T]}(t) \cdot \frac{y}{(x-t)^2 + y^2}\, dt \\ &= \frac{1}{\pi} \int_{-T}^{T} \frac{y}{(x-t)^2 + y^2}\, dt. \end{aligned}$$

Now it is a simple matter to rewrite the integral as

$$\begin{aligned} \omega(x + iy, \Omega, E) &= \frac{1}{\pi} \frac{1}{y} \int_{-T}^{T} \frac{1}{\left(\frac{x-t}{y}\right)^2 + 1}\, dt \\ &= -\frac{1}{\pi} \mathrm{Tan}^{-1} \left(\frac{x-t}{y} \right) \Big|_{-T}^{T} \\ &= \frac{1}{\pi} \mathrm{Tan}^{-1} \left(\frac{x+T}{y} \right) - \frac{1}{\pi} \mathrm{Tan}^{-1} \left(\frac{x-T}{y} \right). \end{aligned}$$

The reader may check that this function is harmonic, tends to 1 as (x, y) approaches $E \subseteq \partial\mathcal{U}$, and tends to 0 as (x, y) approaches $\partial\mathcal{U} \setminus E$.

Glancing at Figure 6.1, we see that the value of $\omega(z, \Omega, E)$ is simply α/π, where α is the angle subtended at the point (x, y) by the interval E.

More generally, if E is any bounded, closed interval in the real line, then the harmonic measure $\omega(z, \mathcal{U}, E)$ will be $1/\pi$ times the angle subtended at the point (x, y) by the interval E. But we can

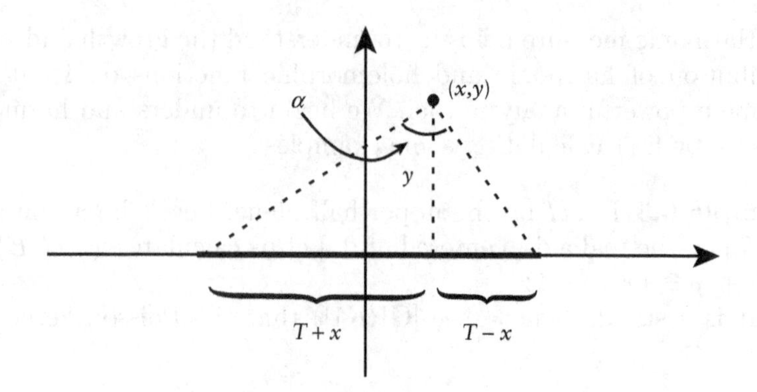

Figure 6.1: The value of $\omega(z, \Omega, E)$.

say more. If now E is the finite disjoint union of closed, bounded intervals,

$$E = I_1 \cup \cdots \cup I_k,$$

then the harmonic measure of E is just the sum of the harmonic measures $\omega(z, \Omega, I_j)$ for each of the individual intervals. So the harmonic measure at z is just $1/\pi$ times the sum of the angles subtended at (x, y) by each of the intervals I_j.

Example 6.6 Let D be the unit disc. Let E be an arc of the circle with central angle α. Then, for $z \in D$,

$$\omega(z, D, E) = \frac{2\theta - \alpha}{2\pi},$$

where θ is the angle subtended by E at z. See Figure 6.2.

It turns out to be convenient to first treat the case where E is the arc from $-i$ to i. See Figure 6.3.

We may simply calculate the Poisson integral of $\chi_{(-\pi/2,\pi/2)}$, where the argument of this characteristic function is the angle in radian measure on the circle. Thus, for $z = re^{i\lambda}$,

$$\begin{aligned} \omega(z, D, E) &= \frac{1}{2\pi} \int_{-\pi/2}^{\pi/2} \frac{1 - r^2}{1 - 2r\cos(\lambda - t) + r^2} \, dt \\ &= \frac{1 - r^2}{2\pi} \int_{-\pi/2}^{\pi/2} \frac{dt}{(1 + r^2) - 2r\cos(\lambda - t)}. \end{aligned}$$

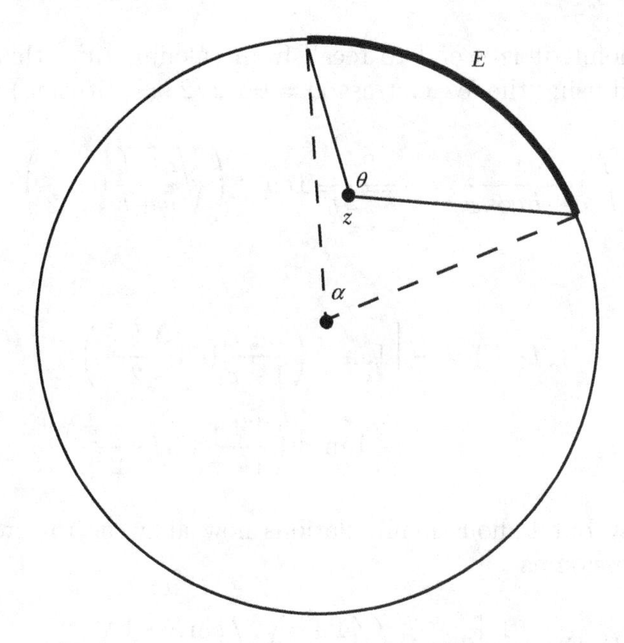

Figure 6.2: The angle subtended by E at z.

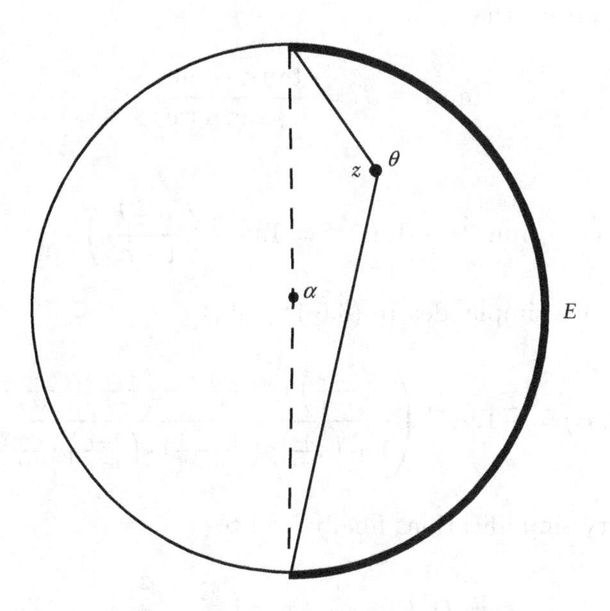

Figure 6.3: The case E is the arc from $-i$ to i.

At this point, it is useful to recall from calculus (and this may be calculated using the Weierstrass $w = \tan x/2$ substitution) that

$$\int \frac{dx}{a + b\cos x} = \frac{2}{\sqrt{a^2 - b^2}} \mathrm{Tan}^{-1}\left(\sqrt{\frac{a-b}{a+b}}\tan\frac{x}{2}\right).$$

Therefore

$$\omega(z, D, E) = \frac{1}{\pi}\left[\mathrm{Tan}^{-1}\left(\frac{1+r}{1-r}\tan\frac{\lambda + \frac{\pi}{2}}{2}\right)\right.$$
$$\left. -\mathrm{Tan}^{-1}\left(\frac{1+r}{1-r}\tan\frac{\lambda - \frac{\pi}{2}}{2}\right)\right].$$

Some easy but tedious manipulations now allow us to rewrite this last expression as

$$\omega(z, D, E) = \frac{1}{\pi}\left[\mathrm{Tan}^{-1}\left(\left(\frac{1+r}{1-r}\right)\cdot\left(\frac{\sin\lambda + 1}{\cos\lambda}\right)\right)\right.$$
$$\left. -\mathrm{Tan}^{-1}\left(\left(\frac{1+r}{1-r}\right)\cdot\left(\frac{\sin\lambda - 1}{\cos\lambda}\right)\right)\right]. \tag{6.6.1}$$

Recall, however, that

$$\tan(\alpha - \beta) = \frac{\tan\alpha - \tan\beta}{1 + \tan\alpha\tan\beta};$$

hence

$$\mathrm{Tan}^{-1}a - \mathrm{Tan}^{-1}b = \mathrm{Tan}^{-1}\left(\frac{a-b}{1+ab}\right). \tag{6.6.2}$$

Applying this simple idea to (6.6.1) yields

$$\omega(z, D, E) = \frac{1}{\pi}\mathrm{Tan}^{-1}\left(\frac{\left(\frac{1+r}{1-r}\right)\left(\frac{\sin\lambda+1}{\cos\lambda}\right) - \left(\frac{1+r}{1-r}\right)\left(\frac{\sin\lambda-1}{\cos\lambda}\right)}{1 + \left(\frac{1+r}{1-r}\right)\left(\frac{\sin\lambda+1}{\cos\lambda}\right)\cdot\left(\frac{1+r}{1-r}\right)\left(\frac{\sin\lambda-1}{\cos\lambda}\right)}\right).$$

Elementary simplifications finally lead to

$$\omega(z, D, E) = \frac{1}{\pi}\mathrm{Tan}^{-1}\left(\frac{1-r^2}{2r\cos\lambda}\right).$$

Now it is helpful to further rewrite the right-hand side of this last expression as

$$\frac{2\left[\mathrm{Tan}^{-1}\left(\frac{1-r\sin\lambda}{r\cos\lambda}\right)+\mathrm{Tan}^{-1}\left(\frac{1+r\sin\lambda}{r\cos\lambda}\right)\right]-\pi}{2\pi}$$

(remembering, of course, that $\pi/2-\mathrm{Tan}^{-1}(\gamma)=\mathrm{Tan}^{-1}(1/\gamma)$). But the preceding expression is just the formula

$$\frac{2\theta-\alpha}{2\pi},$$

for the special angle $\alpha=\pi$, which was enunciated at the start of the example.

For the case of general E, we may first suppose that E is an arc centered at the point $1\in\partial D$. Second, we may reduce the general case to the one just calculated by using a Möbius transformation.

Example 6.7 Suppose that U is an annulus with radii $0<r_1<r_2<\infty$. Then one may verify by inspection that, if E is the outer boundary circle of the annulus, then

$$\omega(z,U,E)=\frac{\log(|z|/r_1)}{\log(r_2/r_1)}.$$

In general, it is a tricky business to calculate exactly the harmonic measure for a given domain U and a given E. But one may often obtain useful estimates. The ensuing discussion will bear out this point.

6.3 Hadamard's Three-Circles Theorem

Capsule: One of the big ideas in modern harmonic analysis is interpolation of operators. If a linear operator T maps a Banach space X to itself and another Banach space Y to itself, then is there automatically a range of "intermediate" spaces on which T is bounded? (This question goes back to M. Riesz's proof of the L^p boundedness of the Hilbert transform in 1926.)

The first real interpolation theorem (which treats the concept of intermediate space) is the Riesz–Thorin theorem. Its proof uses Hadamard's three-circles theorem.

Hadamard's three-circles theorem (sometimes called the "three-lines theorem") is an important sharpening of the classical maximum principle. It has proved useful in various parts of analysis, notably in proving the Riesz–Thorin interpolation theorem for linear operators. Here we shall give a thorough treatment of the three circles theorem from the point of view of harmonic measure. Afterwards we shall discuss the result of Riesz and Thorin.

We begin by treating some general comparison principles regarding harmonic measure. It will facilitate our discussion to first introduce a slightly more general concept of harmonic measure.

Let $U \subseteq \mathbb{C}$ be, as usual, a domain. Let A be a closed set in the extended plane $\widehat{\mathbb{C}}$. Let E denote that part of $\partial(U \setminus A)$ that lies in A. [In what follows, we will speak of pairs (U, A).] Then $\omega(z, U \setminus A, E)$ will be called *the harmonic measure of A with respect to U*, assuming that the geometry is simple enough that we can compute this number component-by-component (of $U \setminus A$). Generally speaking, in the present chapter, we assume that all boundaries that arise consist of finitely many Jordan curves. This standing hypothesis is made so that we can readily apply Lindelöf's Proposition 6.1. Often this standing hypothesis will go unspoken.

The first new comparison tool is called the *majorization principle.*

Prelude: We consider here a transformation property for harmonic measure.

THEOREM 6.8 *Consider two pairs (U, A) and $(\widetilde{U}, \widetilde{A})$. Let*

$$f : U \setminus A \to \widetilde{U}$$

be holomorphic. Assume furthermore that, if $U \ni z \to A$, then $\widetilde{U} \ni f(z) \to \widetilde{A}$. Then

$$\omega(z, U, A) \leq \omega(f(z), \widetilde{U}, \widetilde{A})$$

for $z \in f^{-1}(\widetilde{U} \setminus \widetilde{A})$.

REMARK 6.9 Of course we consider only holomorphic f because we want a mapping that preserves harmonic functions under composition.

Proof of Theorem 6.8: Let us abbreviate

$$\omega = \omega(z, U, A) \qquad \text{and} \qquad \widetilde{\omega} = (f(z), \widetilde{U}, \widetilde{A}).$$

Now we apply the maximum principle to $\omega - \widetilde{\omega}$ on a connected component V of $f^{-1}(\widetilde{U} \setminus \widetilde{A})$. As z approaches the boundary of V, either

z tends to a boundary point of U that is not on A, or else $f(z)$ tends to $\widetilde{A}$. In either of these circumstances, $\limsup_{z \to \partial V}(\omega - \widetilde{\omega}) \leq 0$ *except* when $f(z)$ tends to an endpoint of the boundary arcs of $\widetilde{U} \setminus \widetilde{A}$ that lie on $\widetilde{A}$. Since there are only finitely many such points, Lindelöf's Proposition 6.1 tells us that the maximum principle still remains valid. We therefore conclude that $\omega \leq \widetilde{\omega}$ on all of $f^{-1}(\widetilde{U} \setminus \widetilde{A})$. □

COROLLARY 6.10 *The function $\omega(z, U, A)$ increases if either U increases or A increases. That is to say, if $U \subseteq U^*$ and $A \subseteq A^*$, then*

$$\omega(z, U, A) \leq \omega(z, U^*, A) \qquad \text{and} \qquad \omega(z, U, A) \leq \omega(z, U, A^*).$$

Proof: We prove the first statement and leave the second for the reader.

We apply the theorem with $f : U \setminus A \to U^*$ being the identity mapping. The result is now immediate. □

Now let U^* be an open disc of radius $R > 0$, and let A^* be a smaller closed disc of radius $0 < r < R$. One may see by inspection that

$$\omega(f(z), U^*, A^*) = \frac{\log(R/|f(z)|)}{\log R/r}.$$

We do not yet say what f is, and there is considerable latitude in practice. Nonetheless, we see immediately that the function ω is identically equal to 1 when $f(z) \in \partial A^*$ and identically equal to 0 when $f(z) \in \partial U^*$. Certainly it is harmonic.

Now we have our first result where the quantitative properties of the harmonic measure play a decisive role.

Prelude: This next result is reminiscent of our arguments (below) using the three-circles theorem to establish the Riesz–Thorin theorem.

THEOREM 6.11 *Let f be a holomorphic function on a domain U. Let A be a closed set. If $|f(z)| \leq M$ in U and $|f(z)| \leq m < M$ on A, then, for $0 \leq \theta \leq 1$,*

$$|f(z)| \leq m^\theta M^{1-\theta}$$

at points z where $\omega(z, U, A) \geq \theta$.

REMARK 6.12 This result is of fundamental philosophical importance. It shows how the harmonic measure is a device for interpolating information about the function f.

Proof: Of course $f : U \to D(0,M)$. We take $\widetilde{A} = D(0,m)$, $A = f^{-1}(\widetilde{A})$, and $\widetilde{U} = f(U)$. Then we apply the previous theorem and the remark following. So

$$\omega(z,U,A) \le \omega(f(z),\widetilde{U},\widetilde{A}) = \frac{\log(M/|f(z)|)}{\log M/m}.$$

At a point z for which $\omega(z,U,A) \ge \theta$, we have

$$\theta \le \frac{\log(M/|f(z)|)}{\log M/m}.$$

This inequality is equivalent to the desired conclusion. □

If we take U and A in this last theorem to be annuli, then we can draw an important and precise conclusion. Namely, we have the following version of the three-circles theorem.

Prelude: We have the celebrated Hadamard three-circles theorem, whose uses in analysis are many and varied.

THEOREM 6.13 (HADAMARD) *Let f be a holomorphic function on an annulus $\mathcal{A} = \{z \in \mathbb{C} : c < |z| < C\}$. For $c < \tau < C$, we let*

$$M(\tau) \equiv \max_{|z|=\tau} |f(z)|.$$

If $c < r < \rho < R < C$, then we have

$$\log M(\rho) \le \frac{\log R - \log \rho}{\log R - \log r} \cdot \log M(r) + \frac{\log \rho - \log r}{\log R - \log r} \cdot \log M(R).$$

(This inequality says quite plainly that $\log M(s)$ is a convex function of $\log s$.)

Proof: Let

$$U = \{z \in \mathbb{C} : r < |z| < R\},$$

and let

$$A = \{z \in \mathbb{C} : r \le |z| \le r + \epsilon\}, \text{ some } \epsilon > 0.$$

We assume that f is holomorphic on U, $|f| \le M$ on U, and $|f| \le m$ on A. The preceding theorem tells us that

$$|f(z)| \le m^{\theta} M^{1-\theta} \tag{6.13.1}$$

on the set where $\omega(z,U,A) \geq \theta$, that is, on the set where

$$\frac{\log(|z|/R)}{\log((r+\epsilon)/R)} \geq \theta.$$

(Note that the roles of $r+\epsilon$ and R are reversed from their occurrence in Example 6.7, just because now we are creating harmonic measure for the inside circle of the annulus.)

In particular, inequality (6.13.1) holds when $|z| = (r+\epsilon)^\theta R^{1-\theta}$. We take

$$\theta = \frac{\log R - \log \rho}{\log R - \log(r+\epsilon)}$$

and

$$|z| = (r+\epsilon)^\theta R^{1-\theta} \equiv \rho_\epsilon.$$

Then (6.13.1) translates to

$$M(\rho_\epsilon) \leq M(r+\epsilon)^\theta \cdot M(R)^{1-\theta}.$$

Letting $\epsilon \to 0^+$ and taking the logarithm of both sides yields the desired result. □

We note that Ahlfors [AHL2], from which our exposition derives, likes to express the conclusion of this last result as

$$\det \begin{pmatrix} 1 & 1 & 1 \\ \log r & \log \rho & \log R \\ \log M(r) & \log M(\rho) & \log M(R) \end{pmatrix} \geq 0.$$

We close this section by formulating the three-lines version of Hadamard's theorem. It is proved from Theorem 6.13 simply with conformal mapping, and we leave the details to the interested reader.

Prelude: We have the "three-lines" version of the Hadamard theorem.

THEOREM 6.14 *Let f be a holomorphic function on the strip $S = \{z \in \mathbb{C} : 0 < \operatorname{Re} z < 1\}$. For $0 < x < 1$ we let*

$$M(x) \equiv \max_{\operatorname{Re} z = x} |f(z)|.$$

If $0 < a < \rho < b < 1$, then we have

$$\log M(\rho) \leq \frac{\log b - \log \rho}{\log b - \log a} \cdot \log M(a) + \frac{\log \rho - \log a}{\log b - \log a} \cdot \log M(b).$$

(This inequality says quite plainly that $\log M(s)$ is a convex function of $\log s$.)

6.4 A Discussion of Interpolation of Linear Operators

Capsule: As predicted in the last **Capsule**, we now apply the three-lines theorem to prove a result about interpolation of operators. This will be a new idea for you, and a profound one.

Marcel Riesz discovered the idea of interpolating linear operators in the following context. Perhaps the most important linear operator in all of analysis is the *Hilbert transform*

$$H : f \mapsto \int_{\mathbb{R}} \frac{f(t)}{x-t}\, dt.$$

This operator arises naturally in the study of the boundary behavior of conjugates of harmonic functions on the disc, in the existence and regularity theory for the Laplacian, in the summability theory of Fourier series, and, more generally, in the theory of singular integral operators. It had been an open problem for some time to show that H is a bounded operator on L^p, $1 < p < \infty$.

It turns out that the boundedness on L^2 is easy to prove and follows from Plancherel's theorem in Fourier analysis. Riesz cooked up some extremely clever tricks to derive boundedness on L^p when p is an even integer. (These ideas are explained in detail in [KRA3].) He needed to find some way to derive therefrom the boundedness on the "intermediate" L^p spaces. Now we introduce enough language to explain precisely what the concept of "intermediate space" means.

Let X_0, X_1, Y_0, Y_1 be Banach spaces. Intuitively, we will think of a linear operator T such that

$$T : X_0 \to Y_0$$

continuously and

$$T : X_1 \to Y_1$$

continuously. We wish to posit the existence of spaces X_θ and Y_θ, $0 \le \theta \le 1$ such that these new spaces are natural "intermediaries" of X_0, X_1 and Y_0, Y_1 respectively. Furthermore, we want that

$$T : X_\theta \to Y_\theta$$

continuously. There is interest in knowing how the norm of T acting on X_θ depends on the norms of T acting on X_0 and X_1.

One rigorous method for approaching this situation is as follows (see [BEL], [STW], and [KAT] for our inspiration). Let X_0, Y_0, X_1, Y_1 be given as in the last paragraph. Suppose that

$$T : X_0 \cap X_1 \longrightarrow Y_0 \cup Y_1$$

is a linear operator with the properties that

(i) $\|Tx\|_{Y_0} \leq C_0 \|x\|_{X_0}$ for all $x \in X_0 \cap X_1$;

(ii) $\|Tx\|_{Y_1} \leq C_0 \|x\|_{X_1}$ for all $x \in X_0 \cap X_1$.

Then we want to show that there is a collection of norms $\| \cdot \|_{X_\theta}$ and $\| \cdot \|_{Y_\theta}$, $0 < \theta < 1$, such that

$$\|Tx\|_{Y_\theta} \leq C_\theta \|x\|_{X_\theta}.$$

So the problem comes down to how to construct the "intermediate norms" $\| \cdot \|_{X_\theta}$ and $\| \cdot \|_{Y_\theta}$ from the given norms $\| \cdot \|_{X_0}$, $\| \cdot \|_{Y_0}$, $\| \cdot \|_{X_1}$, and $\| \cdot \|_{Y_1}$.

There are in fact a number of paradigms for effecting the indicated construction. Most prominent among these are the "real method" (usually attributed to Lyons and Peetre—see [LYP]) and the "complex method" (usually attributed to Calderón—see [CAL]). In this text we shall concentrate on the complex method, which was inspired by the ideas of Riesz and Hadamard that were described at the beginning of this section.

Instead of considering the general paradigm for complex interpolation, we shall concentrate on the special case that was of interest to Riesz and Thorin. We now formulate our main theorem.

Prelude: The Riesz–Thorin theorem was the very first theorem in the study of interpolation of operators. This is now a very well-developed part of analysis, and there are entire monographs devoted to the subject.

THEOREM 6.15 *Let $1 \leq p_0 < p_1 \leq \infty$ and $1 \leq q_0 < q_1 \leq \infty$. Suppose that*

$$T : L^{p_0}(\mathbb{R}^2) \cap L^{p_1}(\mathbb{R}^2) \longrightarrow L^{q_0}(\mathbb{R}^2) \cup L^{q_1}(\mathbb{R}^2)$$

is a linear operator on Lebesgue spaces satisfying

$$\|Tf\|_{L^{q_0}} \leq C_0 \|f\|_{L^{p_0}}$$

and

$$\|Tf\|_{L^{q_1}} \leq C_0 \|f\|_{L^{p_1}}.$$

Define p_θ and q_θ by

$$\frac{1}{p_\theta} = (1-\theta)\cdot\frac{1}{p_0} + \theta\cdot\frac{1}{p_1}$$

and

$$\frac{1}{q_\theta} = (1-\theta)\cdot\frac{1}{q_0} + \theta\cdot\frac{1}{q_1},$$

with $0 \le \theta \le 1$. Then T is a bounded operator from the L^{p_θ} norm to the L^{q_θ} norm, and

$$\|Tf\|_{L^{q_\theta}} \le C_0^{1-\theta}C_1^\theta \|f\|_{L^{p_\theta}}$$

for all $f \in L^{p_0} \cap L^{p_1}$.

Proof: Fix a nonzero function f that is continuous and with compact support in $\mathbb{R}^2$. We will prove an a priori inequality for this f, and then extend to general f at the end. Now certainly $f \in L^{p_0} \cap L^{p_1}$. We consider the holomorphic function

$$H : \zeta \mapsto \frac{\left(\int_{\mathbb{R}^2} |Tf(x)|^{(1-\zeta)q_0+\zeta q_1}\,dx\right)^{(1-\zeta)/q_0+\zeta/q_1}}{\left(\int_{\mathbb{R}^2} |f(x)|^{(1-\zeta)p_0+\zeta p_1}\,dx\right)^{(1-\zeta)/p_0+\zeta/p_1}}$$

on the strip $S = \{\zeta \in \mathbb{C} : 0 < \mathrm{Re}\,\zeta < 1\}$. We define

$$M(s) = \sup_{\mathrm{Re}\,\zeta = s} |H(\zeta)|,\ 0 < s < 1.$$

Notice that

$$\begin{aligned} M(0) &= \sup_{t\in\mathbb{R}} \left| \frac{\int_{\mathbb{R}^2} |Tf(x)|^{(1-it)q_0+itq_1}\,dx^{(1-it)/q_0+it/q_1}}{\int_{\mathbb{R}^2} |f(x)|^{(1-it)p_0+itp_1}\,dx^{(1-it)/p_0+it/p_1}} \right| \\ &\le \sup_{t\in\mathbb{R}} \frac{\int_{\mathbb{R}^2} |Tf(x)|^{q_0}\,dx^{1/q_0}}{\int_{\mathbb{R}^2} |f(x)|^{p_0}\,dx^{1/p_0}} \\ &\le C_0. \end{aligned}$$

A similar calculation shows that

$$M(1) \le C_1.$$

Obviously this is grist for the three-lines theorem. We may conclude that $M(s) \le C_0^{1-s}\cdot C_1^s$, $0 < s < 1$.

Now let p_ζ, q_ζ be given by

$$\frac{1}{p_\zeta} = (1-\zeta) \cdot \frac{1}{p_0} + \zeta \cdot \frac{1}{p_1}$$

and

$$\frac{1}{q_\zeta} = (1-\zeta) \cdot \frac{1}{q_0} + \zeta \cdot \frac{1}{q_1}.$$

Let $\zeta = s + it$. Then

$$\frac{\|Tf\|_{L^{q_\zeta}}}{\|f\|_{L^{p_\zeta}}} \leq M(s) \leq C_0^{1-s} C_1^s.$$

In other words,

$$\|Tf\|_{L^{q_s}} \leq C_0^{1-s} C_1^s \|f\|_{L^{p_s}}.$$

This is the desired conclusion for a function f that is continuous with compact support.

The result for general $f \in L^{p_\theta}$ can be achieved with a simple approximation argument. □

6.5 The F. and M. Riesz Theorem

Capsule: One of the big ideas of twentieth century function theory is that due to the brothers F. and M. Riesz. The result has many interpretations—in terms of the absolute continuity of harmonic measure, in terms of the vanishing of negative Fourier coefficients, and many other ideas as well.

One of the classic results of function theory was proved by the brothers F. and M. Riesz. It makes an important statement about the absolute continuity of harmonic measure. We begin with a preliminary result that captures the essence of the theorem. Our exposition here follows the lead of [GARM, Ch. 6].

For $0 < p < \infty$ we define the *Hardy space*

$$H^p(D) = \left\{ f \text{ holomorphic on } D : \sup_{0<r<1} \left(\int_0^{2\pi} |f(re^{i\theta})|^p \, d\theta \right)^{1/p} \right.$$
$$\left. \equiv \|f\|_{H^p} < \infty \right\}.$$

Hardy spaces are generalizations of the more classical space $H^\infty(D)$ of bounded, holomorphic functions.

The most important fact about a function in a Hardy space is that it has a boundary function. We enunciate that result here, and refer the reader to [KRA2, Chapter 8] for the details.

Prelude: The theory of boundary limits for functions in Hardy spaces goes back to work of Fatou in 1906. Later on, Riesz and Hardy played a decisive role in fleshing out the theory and pointing in the directions that we study today.

THEOREM 6.16 *Let $0 < p \leq \infty$ and let $f \in H^p(D)$. Then there is a function $f^* \in L^p(\partial D)$ such that:*

(6.16.1) $\lim_{r \to 1^-} f(re^{i\theta}) = f^*(e^{i\theta})$ *for almost every $\theta \in [0, 2\pi)$.*

(6.16.2) *Let $f_r(e^{i\theta}) = f(re^{i\theta})$ for $0 < r < 1$. Assume that $0 < p < \infty$. Then $\lim_{r \to 1^-} \|f - f_r\|_{L^p(\partial D)} = 0$.*

In the following discussions we shall consider rectifiable curves. Basically, a rectifiable curve is a curve with finite length. The technical definition is that the curve is locally the Lipschitz image of the unit interval (see [FED]). But context will make it clear that the intuitive notion of rectifiability will suffice for our purposes.

Prelude: Here it is important that we relate rectifiability for a curve with the theory of Hardy spaces. Hausdorff measure also comes into play.

THEOREM 6.17 *Let U be a domain such that $\gamma = \partial U$ is a Jordan curve. Let*

$$\varphi : D \to U$$

be a conformal map. Then the curve γ is rectifiable if and only if $\varphi' \in H^1$. In case $\varphi' \in H^1$, then we have

$$\|\varphi'\|_{H^1} = \text{length}(\gamma) = \mathcal{H}^1(\gamma). \tag{6.17.1}$$

Here $\mathcal{H}^1$ is one-dimensional Hausdorff measure.

REMARK 6.18 It is known (see [KRA2, Chapter 8]) that a function in H^1 has a boundary limit function that is in $L^1(\partial D)$. Thus the hypothesis of the theorem says, essentially, that φ is absolutely continuous on the boundary of D. It makes sense, then, that φ would preserve length on the boundary. Theorem 6.17 is essentially equivalent to the result of F. and M. Riesz that we described in Remark 6.4.

Proof of Theorem 6.17: Once again, we invoke Carathéodory's theorem; thus we know that φ and its inverse extend continuously and univalently to their respective boundaries. Let us assume that $\varphi' \in H^1$. Let

$$0 = \theta_0 < \theta_1 < \theta_2 < \cdots < \theta_k = 2\pi$$

be a partition of $[0, 2\pi]$. Then of course

$$\begin{aligned}\sum_{j=1}^{k} \left|\varphi(e^{i\theta_j}) - \varphi(e^{i\theta_{j-1}})\right| &= \lim_{r \to 1^-} \sum_{j=1}^{k} \left|\varphi(re^{i\theta_j}) - \varphi(re^{i\theta_{j-1}})\right| \\ &= \lim_{r \to 1^-} \sum_{j=1}^{k} \left| \int_{\theta_{j-1}}^{\theta_j} \varphi'(re^{i\theta}) ire^{i\theta}\, d\theta \right| \\ &\leq \|\varphi'\|_{H^1}. \qquad (6.17.2)\end{aligned}$$

But plainly the length of γ is the supremum, over all partitions of $[0, 2\pi)$, of the left-hand side of (6.17.2). We conclude that γ is rectifiable and

$$\text{length}(\gamma) \leq \|\varphi'\|_{H^1}.$$

For the converse, assume that γ is rectifiable. If $0 < r < 1$ is fixed, then let $\gamma_r = \varphi(\{z \in \mathbb{C} : |z| = r\})$. Let $\epsilon > 0$. Now choose a partition $\{\theta_0, \theta_1, \ldots, \theta_k\}$ of the interval $[0, 2\pi)$ as before so that

$$\sum_{j=1}^{k} \left|\varphi(re^{i\theta_j}) - \varphi(re^{i\theta_{j-1}})\right| \geq \text{length}(\gamma_r) - \epsilon.$$

We write

$$\eta(z) = \sum_{j=1}^{k} \left|\varphi(ze^{i\theta_j}) - \varphi(ze^{i\theta_{j-1}})\right|.$$

Then η, being the sum of absolute values of holomorphic functions, is subharmonic. By Carathéodory's theorem, η is continuous on $\overline{D}$. Hence

$$\sup_D \eta(z) = \sup_\theta \eta(e^{i\theta}) \leq \text{length}(\gamma).$$

We conclude that

$$\int_0^{2\pi} |\varphi'(re^{i\theta})|\, d\theta = \text{length}(\gamma_r) \leq \eta(r) + \epsilon \leq \text{length}(\gamma) + \epsilon.$$

It follows that $\phi' \in H^1$ and equality (6.17.1) is valid. We conclude by noting that, for a rectifiable curve, the ordinary notion of length and the one-dimensional Hausdorff measure are the same. □

The reason that we include Theorem 6.17 in the present chapter—apart from its general aesthetic interest—is that it can be interpreted in the language of harmonic measure. Let notation be as in Theorem 6.17. To wit, let $\gamma = \partial U$ be a rectifiable Jordan curve as above and let $F \subseteq \gamma$ be a subcurve. Let φ be a conformal mapping of U to the unit disc D. Let $F = \varphi(E)$ and $\alpha = \varphi(0)$. Then Carathéodory's theorem, Example 6.6, and Hadamard's theorem tell us that

$$\omega(\alpha, U, F) = \omega(0, D, E) = \frac{1}{2\pi}|E|,$$

where the $|\cdot|$ denotes arc length. Since F is an arc, we see that Theorem 6.17 and the proof of (6.17.1) tell us that

$$\mathcal{H}^1(F) = \mathcal{H}^1(\varphi(E)) = \lim_{r\to 1^-} \int_E |\varphi'(re^{i\theta})|\, d\theta = \int_E |\varphi'(e^{i\theta})|\, d\theta. \quad (6.18)$$

Of course if (6.18) holds for arcs, then, by passing to unions and intersections, we see that it holds for Borel sets F. Thus we derive the important conclusion

$$\omega(\alpha, U, F) = 0 \Rightarrow \mathcal{H}^1(F) = 0.$$

Conversely, since (by a standard uniqueness theorem—see [KOO])

$$|\{\theta : |\varphi'(e^{i\theta})| = 0\}| = 0,$$

we see that

$$\mathcal{H}^1(F) = 0 \Rightarrow \omega(\alpha, U, F) = 0.$$

What we have proved, then, is that when $\gamma = \partial U$ is rectifiable then harmonic measure for U and linear measure on γ are mutually absolutely continuous. We now summarize this result in a formally enunciated theorem.

Prelude: The F. and M. Riesz theorem is one of the celebrated results of classical function theory. There are many different renditions of the result, and many different proofs. We give here a geometric rendition of the theorem.

THEOREM 6.19 (F. AND M. RIESZ, 1916) *Let U be a simply connected planar domain such that $\gamma = \partial U$ is a rectifiable Jordan curve.*

Assume that

$$\varphi : D \to U$$

is conformal. Then $\varphi' \in L^1(\partial D)$. *For any Borel set* $E \subseteq \partial D$ *it holds that*

$$\mathcal{H}^1(\varphi(E)) = \int_E |\varphi'(e^{i\theta}| \, d\theta.$$

Also, for any Borel set $F \subseteq \partial U$ *and any point* $\alpha \in U$,

$$\omega(\alpha, U, F) = 0 \Leftrightarrow \mathcal{H}^1(F) = 0.$$

We conclude by noting that Theorems 6.17 and 6.19 are equivalent, just because formula (6.17.1) is valid precisely when $\varphi' \in H^1$.

Chapter 7

Extremal Length

Prologue: The idea of extremal length was developed by L. Ahlfors and A. Beurling in the 1940s. It is an important conformal invariant that measures the "shape" of a domain from the complex-analytic point of view.

Calculations of extremal length are tricky, but tend to be informative. Extremal length is a natural tool for proving, for instance, that annuli of different radius ratios are conformally inequivalent.

The idea of quasiconformal mapping (a generalization of conformal mapping) interacts naturally with extremal length.

One usually calculates the extremal length of a collection of curves in a domain. By choosing these curves propitiously, one can elicit concrete geometric information about the domain.

We give here the basics of extremal length theory.

The concept of extremal length, first developed by Ahlfors and Beurling, is a useful conformal invariant. It is a decisive tool in the study of both conformal and quasiconformal mappings.

Let $U, U' \subseteq \mathbb{C}$ be domains and Γ a collection of curves in U. If $\Phi : U \to U'$ is a conformal mapping, then we can consider the image of Γ under Φ. It turns out (as we shall see in detail below) that Γ and its image have the same extremal length.

7.1 Some Definitions

Capsule: Extremal length is a relatively new conformal invariant that is a useful way of measuring the geometry of a domain in conformally invariant language.

Fix a domain $U \subseteq \mathbb{C}$. Also let $\rho : U \to [0, \infty]$ be Borel measurable. If $\gamma : I \to U$ is any rectifiable curve in U then we let

$$L_\rho(\gamma) \equiv \int_\gamma \rho |dz|.$$

Here $|dz|$ is the Euclidean element of length. We call the quantity $L_\rho(\gamma)$ the ρ-length of γ. It is possible that $L_\rho(\gamma) = +\infty$.

Put in other words, the value of $L_\rho(\gamma)$ is precisely equal to the integral of the Borel measurable function $\rho(\gamma(t))$ with respect to the Borel measure on I for which the measure of every subinterval $J \subseteq I$ is the length of the restriction of γ to J.

Let Γ be a collection of rectifiable curves in U. Now define

$$L_\rho(\Gamma) = \inf_{\gamma \in \Gamma} L_\rho(\gamma).$$

The *area* of ρ is defined to be

$$A(\rho) = \iint_U \rho^2 \, dx dy.$$

Finally, the *extremal length* of Γ is

$$\lambda_U(\Gamma) \equiv \sup_\rho \frac{L_\rho(\Gamma)^2}{A(\rho)}.$$

Here the supremum is over all Borel measurable $\rho : U \to [0, \infty]$ with $0 < A(\rho) < \infty$.

We also define the *modulus* of Γ to be $1/\lambda_U(\Gamma)$.

The *extremal distance* between two subsets A and B of U is defined to be the extremal length of the collection of curves in U with one endpoint in A and the other endpoint in B.

The conformal invariance of extremal length follows immediately from the change of variables formula in the two integrals. See the next section.

7.2 The Conformal Invariance of Extremal Length

Capsule: The main reason that extremal length is important is that it is a conformal invariant. We develop that idea here.

Let $\Phi : U \to \widehat{U}$ be a conformal mapping (that is to say, a one-to-one, onto, invertible holomorphic mapping) of planar domains. Let Γ be a collection of rectifiable curves in U, and let $\widehat{\Gamma} \equiv \{\Phi \circ \gamma : \gamma \in \Gamma\}$ be the image curves under Φ. We claim that $\lambda_U(\Gamma) = \lambda_{\widehat{U}}(\widehat{\Gamma})$.

For the proof, suppose that $\widehat{\rho} : \widehat{U} \to [0, \infty]$ is Borel measurable. Define

$$\rho(z) = |\Phi'(x)| \cdot \widehat{\rho}(\Phi(z)).$$

The change of variable $w = \Phi(z)$ gives

$$\begin{aligned} A(\rho) &= \int_U \rho(z)^2 \, dz d\overline{z} \\ &= \int_U \widehat{\rho}(\Phi(z))^2 |\Phi'(z)|^2 \, dz d\overline{z} \\ &= \int_{\widehat{U}} \widehat{\rho}(w)^2 \, dw d\overline{w} \\ &= A(\widehat{\rho}). \end{aligned}$$

Now let $\gamma \in \Gamma$. Set $\widehat{\gamma} = \Phi \circ \gamma$. We again use a change of variable to calculate

$$L_\rho(\gamma) = \int_\gamma \widehat{\rho}(\Phi(z)) |\Phi'(z)| \, |dz| = \int_{\widehat{\gamma}} \rho(w) \, |dw| = L_{\widehat{\rho}}(\widehat{\gamma}).$$

We may immediately conclude that

$$\lambda_U(\Gamma) \geq \lambda_{\widehat{U}}(\widehat{\Gamma}).$$

But the same argument applies to the inverse mapping Φ^{-1}. That gives equality, and ends the proof.

7.3 Some Examples

Capsule: Some good examples help to illustrate the utility and significance of extremal length.

We now do some calculations to exhibit extremal length explicitly.

Example 7.1 Let

$$\mathcal{R} = \{(x, y) : 0 < x < w, 0 < y < h\}$$

be a rectangle in the complex plane. Let Γ be the set of all finite length, rectifiable curves which begin on the left edge of the rectangle and end on the right edge of the rectangle. We shall now demonstrate that

$$\lambda_{\mathcal{R}}(\Gamma) = \frac{w}{h}.$$

First of all, we may take $\rho \equiv 1$ on $\mathcal{R}$. Then we see that $A(\rho) = wh$ and $L_\rho(\gamma) = w$ for any $\gamma \in \Gamma$. The definition of $\lambda_{\mathcal{R}}(\Gamma)$ as a supremum then tells us that

$$\lambda_{\mathcal{R}}(\Gamma) \geq \frac{w^2}{wh} = \frac{w}{h}.$$

For the opposite inequality, consider an arbitrary Borel measurable $\rho : \mathcal{R} \to [0, \infty]$ such that $\ell \equiv L_\rho(\Gamma) > 0$. If $y \in (0, h)$, then we let $\gamma_y(t) = iy + wt$. Then $A(\gamma_y) \in \Gamma$, so that $\ell \leq L_\rho(\gamma_y)$. We may write this last inequality as

$$\ell \leq \int_0^1 \rho(iy + wt) w \, dt.$$

Now integrating this inequality over $y \in (0, h)$ gives

$$h\ell \leq \int_0^h \int_0^w \rho(iy + wt) w \, dt dy.$$

We perform the change of variable $x = wt$ and then apply the Cauchy-Schwarz inequality to obtain

$$\begin{aligned} h\ell &\leq \int_0^h \int_0^w \rho(x + iy) \, dx dy \\ &\leq \left(\int_{\mathcal{R}} \rho^2 \, dx dy \int_{\mathcal{R}} dx dy \right)^{1/2} \\ &= (whA(\rho))^{1/2}. \end{aligned}$$

This yields that

$$\lambda_{\mathcal{R}}(\Gamma) \leq \frac{w}{h},$$

as desired.

In fact, the proof presented above shows that the extremal length of Γ is equal to the extremal length of the smaller collection of curves given by $\{\gamma_y : y \in (0,h)\}$.

It can be observed that the extremal length of the family of curves Γ' that connect the bottom edge of $\mathcal{R}$ to the top edge of $\mathcal{R}$ satisfies $\lambda_{\mathcal{R}}(\Gamma') = h/w$—just using the same arguments. Therefore

$$\lambda_{\mathcal{R}}(\Gamma) \cdot \lambda_{\mathcal{R}}(\Gamma') = 1.$$

It is reasonable to think of this equation as a duality property of extremal length. And this duality property can be useful.

For, in general, obtaining a lower bound for $\lambda_U(\Gamma)$ is easier than obtaining an upper bound—just because the lower bound involves choosing a propitious ρ and then estimating $L_\rho(\Gamma)^2/A(\rho)$ while the upper bound necessitates proving an estimate for all possible ρ. Duality, on the other hand, enables us to translate a lower bound on $\lambda_U(\Gamma')$ into an upper bound on $\lambda_U(\Gamma)$.

As Ahlfors points out in [AHL1, p. 52], the extremal length depends on the family of curves Γ but *not* on the domain. To see this, suppose that $U \subseteq U'$ are planar domains. Given ρ on U, we define

$$\rho'(z) = \begin{cases} \rho(z) & \text{if} \quad z \in U \\ 0 & \text{if} \quad z \in U' \setminus U. \end{cases}$$

Then ρ' is certainly Borel measurable, and

$$L_\rho(\Gamma) = L_{\rho'}(\Gamma) \qquad \text{and} \qquad A(\rho) = A(\rho').$$

Thus

$$\lambda_{U'}(\Gamma) \geq \lambda_U(\Gamma).$$

For the opposite inequality, begin with a ρ' on U' and let ρ be the restriction of ρ' to U. We see immediately that $\lambda_U(\Gamma)$ is the same for any open set U that contains all the curves γ in Γ.

As a result of the discussion in the last two paragraphs, we will henceforth write $\lambda(\Gamma)$ rather than $\lambda_U(\Gamma)$.

Example 7.2 Now let us examine the extremal length in an annulus. Fix $0 < r_1 < r_2 < \infty$. Define

$$\mathcal{A} = \{z \in \mathbb{C} : r_1 < |z| < r_2\}.$$

Further let $C_1 = \{z \in \mathbb{C} : |z| = r_1\}$ and $C_2 = \{z \in \mathbb{C} : |z| = r_2\}$. We consider the extremal length between C_1 and C_2. This is, by definition, the extremal length of the collection Γ of curves connecting C_1 to C_2.

To find a lower bound on $\lambda(\Gamma)$, we take $\rho(z) = 1/|z|$. If γ is a curve from C_1 to C_2, we see that

$$\int_\gamma |z|^{-1} ds \geq \int_\gamma |z|^{-1} d|z| = \int_\gamma d \log |z| = \log(r_2/r_1).$$

On the other hand,

$$A(\rho) = \int_A |z|^{-2} \, dxdy = \int_0^{2\pi} \int_{r_1}^{r_2} r^{-2} r \, dr d\theta = 2\pi \log(r_2/r_2).$$

We conclude that

$$\lambda(\Gamma) \geq \frac{\log(r_2/r2)}{2\pi}.$$

Now we shall show that this inequality is really an equality. Consider an arbitrary Borel measurable ρ such that $\ell = L_\rho(\Gamma) > 0$. For $\theta \in [0, 2\pi)$, we let $\gamma_\theta : (r_1, r_2) \to \mathcal{A}$ denote the curve $\gamma_\theta(r) = e^{i\theta} r$. Then

$$\ell \leq \int_{\gamma_\theta} \rho \, ds = \int_{r_1}^{r_2} \rho(e^{i\theta} r) \, dr.$$

We integrate in θ and apply the Cauchy-Schwarz inequality to obtain

$$2\pi\ell \leq \int_{\mathcal{A}} \rho \, dr d\theta \leq \left(\int_{\mathcal{A}} \rho^2 r \, dr d\theta \right)^{1/2} \cdot \left(\int_0^{2\pi} \int_{r_1}^{r_2} \frac{1}{r} \, dr d\theta \right)^{1/2}.$$

Squaring both sides gives

$$4\pi^2 \ell^2 \leq A(\rho) \cdot 2\pi \cdot \log(r_2/r_1).$$

This inequality implies the upper bound $\lambda(\Gamma) \leq (2\pi)^{-1} \log(r_2/r_1)$. When put together with the lower bound, we obtain the final result

$$\lambda(\Gamma) = \frac{\log(r_2/r_1)}{2\pi}.$$

A nice corollary of this last example is that, if $\mathcal{A} = \{z \in \mathbb{C} : r_1 < |z| < r_2\}$ and $\mathcal{B} = \{z \in \mathbb{C} : s_1 < |z| < s_2\}$, and if $r_2/r_1 \neq s_2/s_1$, then $\mathcal{A}$ and $\mathcal{B}$ cannot be conformally equivalent.

Chapter 8

Analytic Capacity

Prologue: The idea of analytic capacity is an old one. It was formally introduced by L. Ahlfors in the 1940s in his study of removable sets for bounded, analytic functions. But these ideas were anticipated by work of Painlevé in the nineteenth century.

Analytic capacity is analogous to measure theory, in that it assigns a numerical "size" to a compact set in the plane. But capacity is much trickier to study. For example, it is only recently that it was proved that analytic capacity is subadditive. It is generally challenging to calculate the capacity of any given set.

It is natural to ask how analytic capacity relates to length, to Hausdorff dimension, and to Lebesgue measure. Considerable effort has been devoted to studying all these questions.

Analytic capacity is also a natural language for studying questions in analytic approximation theory.

This chapter introduces the key ideas in the theory of analytic capacity.

It is well known to every student that, if U is a domain (even the entire complex plane) and $P \in U$, then a bounded, holomorphic function on $U \setminus \{P\}$ continues analytically to all of U. This is a classical result of Riemann.[1] It is of interest to know which compact sets

[1]Less well known is the fact that it is sufficient for f to be square integrable.

can replace P and give the same result. In studying this question in 1947, Lars Ahlfors [AHL4] formulated the idea of analytic capacity.

Definition 8.1 A set $K \subseteq \mathbb{C}$ is said to be *removable* if, whenever U is a domain containing K, then every holomorphic function on $U \setminus K$ has a holomorphic continuation to all of U.

These ideas suggest the following question:

> Given a compact set $E \subseteq \mathbb{C}$, when does there exist a non-constant, bounded, holomorphic function f on $\mathbb{C} \setminus E$?

A partial answer to the question was given by Painlevé in the nineteenth century.

Prelude: We provide now one of the key technical ideas in the theory of removable sets. This idea dates back to the nineteenth century.

THEOREM 8.2 (PAINLEVÉ) *Let $E \subseteq \mathbb{C}$ be compact. Assume that, for every $\epsilon > 0$, the set E can be covered by a collection of open discs whose radii do not exceed ϵ. Then the set of bounded, analytic functions on $\mathbb{C} \setminus E$ consists only of the constant functions.*

Proof: For each $\epsilon > 0$, cover E by a collection of discs D_j of radius r_j such that $\sum_j r_j < \epsilon$. Set $D_\epsilon = \cup_j D_j$. Let $\Gamma_\epsilon \equiv \partial D_\epsilon$. Let f be a bounded, holomorphic function on $\mathbb{C} \setminus E$. Now, if $z \notin \overline{D_\epsilon}$, then

$$f'(z) = -\frac{1}{2\pi i} \oint_{\Gamma_\epsilon} \frac{f'(\zeta)}{(z-\zeta)^2}\, d\zeta.$$

Then

$$|f'(z)| \leq \frac{2\pi\epsilon \cdot \sup_{\Gamma_\epsilon} |f'(\zeta)|}{2\pi [d(z, \Gamma_\epsilon)]^2}.$$

As $\epsilon \to 0$, we see that $|f'(z)| = 0$. So f is constant. □

If $U \subseteq \mathbb{C}$ then we let $H^\infty(U)$ be the space of bounded, holomorphic functions on U. Now we have Ahlfors's definition:

Definition 8.3 Let $E \subseteq \mathbb{C}$. Let $U = \mathbb{C} \setminus E$. We define the *analytic capacity* $\gamma(E)$ of E by

$$\gamma(E) = \sup\{|f'(\infty)| : f \in H^\infty(U) \text{ and } \|f\|_{\sup} \leq 1\}.$$

Here $f'(\infty)$ is calculated relative to the local coordinate $z = 1/\zeta$ on the Riemann sphere as

$$f'(\infty) = \lim_{\zeta \to 0} f'(1/\zeta) = \lim_{\zeta \to 0} \frac{f(1/\zeta) - f(\infty)}{\zeta - 1/\zeta} = \lim_{z \to \infty} z \cdot (f(z) - f(\infty)).$$

REMARK 8.4 Use the notation from the definition. Consider the Möbius transformation

$$\varphi(z) = \frac{f(z) - f(\infty)}{1 - \overline{f(\infty)} f(z)}.$$

This function is certainly a bounded, analytic function on $\mathbb{C} \setminus E$. Thus we need only consider functions f with $f(\infty) = 0$ because $\varphi(\infty) = 0$.

Now we have

PROPOSITION 8.5 *These properties hold for the analytic capacity:*

(a) *If $f(\infty) = 0$, then γ has this invariance property:*

$$\gamma(aE + b) = |a|\gamma(E).$$

(b) *If $E \subseteq F$, then $\gamma(E) \leq \gamma(F)$.*

Proof: The first statement follows from the equation

$$\lim_{z \to \infty} z f(az + b) = a f'(\infty).$$

The second statement is obvious. □

The next result is important for calculations.

PROPOSITION 8.6 *Assume that E is connected, simply connected, but not a singleton. Let g be the conformal map of $\widehat{\mathbb{C}} \setminus E$ to the unit disc D satisfying $g(\infty) = 0$, $g'(\infty) > 0$. Then $\gamma(E) = |g'(\infty)|$.*

Proof: Clearly $|g'(\infty)| \leq \gamma(E)$ by the definition of γ. If f is any other candidate mapping for the analytic capacity of E, then apply Schwarz's lemma to $F \equiv f \circ g^{-1}$. We thus obtain $|f'(\infty)| \leq |g'(\infty)|$ for any such f. Thus $\gamma(E) \leq |g'(\infty)|$, completing the proof. □

8.1 Calculating Analytic Capacity

Capsule: Analytic capacity is most meaningful in the context of concrete examples. And there is always some interesting function theory involved in working the examples.

Example 8.7 Let E be the real interval $[-2,2] = \{x + i0 : -2 \leq x \leq 2\}$. Define $g(z) = z + 1/z$. Then g is a conformal mapping onto the complement of E that takes the unit circle to E. Thus $\gamma(E) = (g^{-1})'(\infty)$, so that we have

$$\gamma(E) = \frac{1}{g'(g(\infty))} = \frac{1}{1 - 1/\infty} = 1.$$

More generally, let E be the real interval $[a,b] = \{x + i0 : a \leq x \leq b\}$. Then, using results of the last section, we see that

$$\gamma(E) = \gamma\left(\frac{b-a}{4}[-2,2] + \frac{a+b}{2}\right) = \frac{b-a}{4}\gamma([-2,2]) = \frac{b-a}{4}.$$

It is actually possible to show (we shall omit the details) that, if $E \subseteq \mathbb{R} \subseteq \mathbb{C}$ is *any* Boreal measurable set, then $\gamma(E)$ is 1/4 the measure of E.

Example 8.8 Let E be the disc $\{z \in \mathbb{C} : |z - z_0| \leq r\}$. Then

$$\widehat{\mathbb{C}} \setminus E = \{z \in \mathbb{C} : |z - z_0| > r\},$$

and the mapping

$$z \longmapsto \frac{r}{z - z_0}$$

is a conformal map of $\widehat{\mathbb{C}} \setminus E$ to the disc D.

If we set $g(z) = r/(z - z_0)$, then, using the local coordinate $z = 1/\xi$, we have

$$g(1/\xi) = \frac{r}{1/\xi - z_0} = \frac{r\xi}{1 - z_0\xi}.$$

Now differentiation shows us that

$$g'\left(\frac{1}{\xi}\right) = \frac{r}{1 - z_0\xi} - \frac{r\xi}{(1 - z_0\xi)^2}.$$

Thus, if we send ξ to 0, we obtain

$$g'(\infty) = \lim_{\xi \to 0} g'(1/\xi) = r.$$

We conclude that the analytic capacity of a disc of radius r is r.

8.2 Analytic Capacity and Removability

Capsule: The whole idea of analytic capacity is that it gives a measure of "how removable" a set is. If a set has analytic capacity zero, then it is removable for bounded, holomorphic functions. If it has positive analytic capacity, then it is nonremovable—and the capacity gives a measure of how nonremovable it is.

Ahlfors's original motivation for conceiving of and studying analytic capacity was the characterization of removable sets. In that vein we have the following essential result.

Prelude: This result shows explicitly how to understand analytic capacity in terms of the concept of removable set. It is a fundamental theorem.

THEOREM 8.9 *Let $E \subseteq \mathbb{C}$ be a compact set. Then the following assertions are equivalent:*

(a) $\gamma(E) = 0$.

(b) *Every bounded, holomorphic function $f : \mathbb{C} \setminus E \to \mathbb{C}$ is constant.*

(c) *E is removable for bounded, analytic functions.*

Proof: It is clear that **(b)** implies **(a)**. Also **(c)** implies **(b)** by Liouville's theorem.

For **(a)** $\Rightarrow$ **(b)**, Suppose that there exists a nonconstant, bounded, holomorphic function $f : \mathbb{C} \setminus E \to \mathbb{C}$ with $f(\infty) = 0$ and $f(z_0) \neq 0$ for some $z_0 \in \mathbb{C} \setminus E$. Let us define

$$g(z) = \begin{cases} \frac{f(z)-f(z_0)}{z-z_0} & \text{if} \quad z \in \mathbb{C} \setminus E \text{ and } z \neq z_0 \\ f'(z_0) & \text{if} \quad z = z_0. \end{cases}$$

We see that g is a bounded, holomorphic function on $\mathbb{C} \setminus E$ and that $g'(\infty) = f(z_0) \neq 0$. Thus $\gamma(E) > 0$, and that is a contradiction.

Finally, for **(b)** $\Rightarrow$ **(c)**, suppose that E satisfies **(b)**. We claim that E must be totally disconnected. If not, then the Riemann Mapping theorem tells us that there is a nonconstant, bounded, analytic function $f : \mathbb{C} \setminus E_0 \to D$ for some subset $E_0 \subseteq E$. That would be a contradiction.

Now let U be an open set that contains E. Let f be a bounded, holomorphic function on $U \setminus E$, and fix a point $z_0 \in U \setminus E$. Since E

is totally disconnected, there are two simple closed curves γ_1, γ_2 in $U \setminus E$ so that z_0 is in the domain bounded by γ_1 but not in the domain bounded by γ_2. Then the Cauchy integral formula tells us that

$$\oint_{\gamma_1} \frac{f(z)}{z - z_0}\, dz + \oint_{\gamma_2} \frac{f(z)}{z - z_0}\, dz$$

defines an analytic extension of f to all of U. So **(c)** holds for E. □

Prelude: This is a very basic, and intuitively appealing, result about analytic capacity. It relates capacity to area.

THEOREM 8.10 *Let $E \subseteq \mathbb{C}$ have positive area, and let U be the complement of E in the Riemann sphere. Then there are nonconstant, holomorphic functions on U that extend continuously to the sphere.*

Proof: Define

$$F(z) = \iint_E \frac{d\xi d\eta}{\zeta - z},$$

with $\zeta = \xi + i\eta$. Then it is clear that F is holomorphic on the complement U of E. Since F is the convolution of the locally integrable function $1/\zeta$ and the characteristic function of E, we see that F is continuous on the entire complex plane. Since $\lim_{z\to\infty} F(z) = 0$, we see that F extends continuously to the entire Riemann sphere. Also F is not constant since $\lim_{z\to\infty} zF(z) = -\text{Area}(E)$. That completes the proof. □

Prelude: Again, we relate capacity to area. This is one of the key theorems in the subject.

THEOREM 8.11 *The analytic capacity of E is bounded below by*

$$\frac{1}{2}\sqrt{\frac{\text{Area}(E)}{\pi}}.$$

Proof: Let F be as in the previous result. At infinity, F has the expansion

$$F(z) = -\frac{\text{Area}(E)}{z} + \frac{a_2}{z^2} + \cdots.$$

Define the positive number R by

$$\pi R^2 = \text{Area}(E).$$

Fix z and let Δ be the disc $\{\zeta \in \mathbb{C} : |\zeta - z| \leq R\}$. Then

$$|F(z)| \leq \iint_E \frac{d\xi d\eta}{|\zeta - z|} = \iint_{E\cap\Delta} \frac{d\xi d\eta}{|\zeta - z|} + \iint_{E\setminus\Delta} \frac{d\xi d\eta}{|\zeta - z|}.$$

Since E and Δ have the same area, so do $E \setminus \Delta$ and $\Delta \setminus E$. But the integrand on the right-hand side of the last displayed equation is larger on $\Delta \setminus E$ than it is on $E \setminus \Delta$. So

$$\iint_{E\setminus\Delta} \frac{d\xi d\eta}{|\zeta - z|} \leq \iint_{\Delta\setminus E} \frac{d\xi d\eta}{|\zeta = z|}.$$

Hence

$$F(z) \leq \iint_\Delta \frac{d\xi d\eta}{|\zeta - z|} = 2\pi R.$$

This gives us a bounded function

$$g(z) = \frac{F(z)}{2\pi R}$$

on U such that $\|g\| \leq 1$ and

$$g(z) = \frac{b_1}{z} + \frac{b_2}{z^2} + \cdots,$$

where

$$|b_1| \geq \frac{R}{2} = \frac{1}{2}\sqrt{\frac{\text{Area}(E)}{\pi}}.$$

In other words, we have estimated the analytic capacity of E from below. □

If U is a domain in $\mathbb{C}$, then we let $H^\infty(U)$ denote the space of bounded, analytic functions on U. And we let $A(U)$ denote the elements of $H^\infty(U)$ that possess continuous extensions to the Riemann sphere. If $E \subseteq \mathbb{C}$, and U is the complement of E, then define

$$\alpha(E) = \sup\{|f'(\infty)| : f \in A(U), \|f\| \leq 1\}.$$

It is useful to compare and contrast $\alpha(E)$ with $\gamma(E)$.

Prelude: This is our opportunity to compare the original analytic capacity γ with the new capacity α. The result is elegant and useful.

THEOREM 8.12 *Let $E \subseteq \mathbb{C}$, and let U be the complement of E. The space $H^\infty(U)$ consists only of the constants if and only if $\gamma(E) = 0$. The space $A(U)$ consists only of the constants if and only if $\alpha(E) = 0$.*

Proof: If $\gamma(E) > 0$ then $H^\infty(U)$ contains nonconstant functions. On the other hand, if $H^\infty(U)$ is nontrivial, then there is an $f \in H^\infty(U)$ with $f(\infty) = 0$ and $f(z_0) \neq 0$ for some $z_0 \neq \infty$. Then the function

$$g(z) = \frac{f(z) - f(z_0)}{z_0 - z}$$

is in $H^\infty(U)$ and has derivative $f(z_0)$ at ∞; hence $\gamma(E) > 0$.

The very same argument shows that $\alpha(U)$ is nontrivial if and only if $\alpha(E) > 0$. □

PROPOSITION 8.13 *For any set E,*

$$\alpha(E) \leq \gamma(E) \leq \mathrm{diam}(E).$$

If E is connected, then

$$\gamma(E) \geq \frac{\mathrm{diam}(E)}{4}.$$

Proof: The first set of inequalities follows from monotonicity of the analytic capacity, just because E lies in a disc of radius equal to $\mathrm{diam}(E)$.

For the second inequality, we may assume that E is not a singleton. Let

$$g(z) = \frac{\gamma(E)}{z} + \frac{a_2}{z^2} + \cdots$$

be the Riemann mapping of the unbounded component of $U = \mathbb{C} \setminus E$ onto the unit disc. Fix $z_0 \in E$, and write

$$f(w) = \frac{\gamma(E)}{g^{-1}(w) - z_0}.$$

Then f is univalent on $\{|w| < 1\}$, $f(0) = 0$, and $f'(0) = 1$. Now the Koebe-Bieberbach theorem tells us that the range of f contains the disc $\{|z| < 1/4\}$. Thus, if $z_1 \in E$, then

$$\frac{\gamma(E)}{|z_1 - z_0|} \geq \frac{1}{4}. \qquad \square$$

One conclusion that we may draw from the last proposition is that, if $\gamma(E) = 0$, then E is totally disconnected.

Prelude: Again we relate capacity to removability for sets. The ideas are fundamental and important.

THEOREM 8.14 *Let E be a relatively closed subset of an open set $U \subseteq \mathbb{C}$ and assume that $\gamma(E) = 0$. If $f \in H^\infty(U \setminus E)$, then f has a unique extension in $H^\infty(U)$.*

Proof: The uniqueness is trivial because E is nowhere dense.

Let $z_0 \in E$. By the last proposition, E is totally disconnected. So there is an analytic, simple, closed curve Γ in $U \setminus E$ which encloses z_0. Let U be the domain bounded by Γ.

Using the Cauchy integral formula, we may write

$$f = f_1 + f_2$$

in a neighborhood of Γ, where $f_1 \in H^\infty(U)$ and $f_2 \in H^\infty(\widehat{\mathbb{C}} \setminus (E \cap U))$. Since $\gamma(E \cap U) = 0$, we see that f_2 is constant. Hence f extends analytically to U. $\square$

If $E \subseteq \mathbb{C}$ and $M > 0$, then define

$$A(E, M) = \{f \in H^\infty(U) : \|f\| \leq M, f(\infty) = 0\}.$$

Prelude: Now we are estimating the size of a function in terms of analytic capacity. The theory is becoming richer.

THEOREM 8.15 *Let $E \subseteq \mathbb{C}$. Let $f \in A(E, 1)$. Then, for $z_0 \in U = \mathbb{C} \setminus E$,*

$$|f(z_0)| \leq \frac{\gamma(E)}{\text{dist}(z_0, E)}.$$

If f has a zero of multiplicity m at ∞, and E has diameter d, then

$$|f(z_0)| \leq \frac{d^{m-1}\gamma(E)}{(\text{dist}(z_0, E))^m}.$$

Proof: Consider the function

$$g(z) = \frac{1}{z - z_0} \cdot \frac{f(z) - f(z_0)}{1 - \overline{f(z_0)} f(z)}.$$

Then $g \in H^\infty(U)$, $g(\infty) = 0$, and

$$|g(z)| \le \frac{1}{\mathrm{dist}(z_0, E)}.$$

Then

$$|f(z_0)| = |g'(\infty)| \le \frac{\gamma(E)}{\mathrm{dist}(z_0, E)}.$$

If f has order m at ∞ and $z_1 \in E$ with $|z_1 - z_0| = \mathrm{dist}(z_0, E)$, then

$$h(z) \equiv \left(\frac{z - z_0}{d}\right)^{m-1} f(z)$$

is in $A(E, 1)$. Hence

$$|f(z_0)| \le \left(\frac{d}{|z - z_0|}\right)^{m-1} \cdot |h(z_0)| \le \frac{d^{m-1}\gamma(E)}{(\mathrm{dist}(z_0, E))^m}. \qquad \square$$

Prelude: Here we learn more about the structure of an extremal function. We are developing in new directions.

THEOREM 8.16 *Let U be a domain such that $\infty \in U$. Assume that ∂U consists of pairwise disjoint analytic Jordan curves $\Gamma_1, \Gamma_2, \ldots, \Gamma_k$. Let $E = \widehat{\mathbb{C}} \setminus U$. Then*

(a) $\alpha(E) = \gamma(E)$.

If $f \in A(E, 1)$ is an extremal function, then

(b) *f is analytic across ∂U.*

(c) *$|f| = 1$ on ∂U.*

(d) *f has k zeros on U.*

Proof: Let σ be a measure on ∂U of measure norm $\|\sigma\| = \alpha(E)$ and so that

$$\int f\, d\sigma = f'(\infty)$$

for all $f \in A(U)$. The measure

$$\mu = \frac{dz}{2\pi i} - \sigma$$

is orthogonal to $A(U)$. Hence, by the F. and M. Riesz theorem,

$$\sigma = \psi(z)\, dz$$

for

$$\psi(z) = \frac{1}{2\pi i} + \frac{b_2}{z^2} + \cdots \in H^1(U).$$

The function ψ solves the dual extremal problem

$$\int_{\partial U} |\psi(\zeta)|\, ds = \inf\left\{ \int_{\partial U} |h(\zeta)|\, ds : h \in H^1(U), h(\infty) = \frac{1}{2\pi} \right\}.$$

We call ψ the *Garabedian function.*

If $g \in A(E,1)$, then $g \cdot \psi \in H^1(U)$ and $|g\psi| \leq |\psi|$ on ∂U. Thus

$$|g'(\infty)| = \left| \int_{\partial U} g(\zeta)\psi(\zeta)\, d\zeta \right| \leq \int_{\partial U} |\psi|\, ds = \alpha(E).$$

Taking $g'(\infty) = \gamma(E)$, we now obtain $\gamma(E) \leq \alpha(E)$, so that **(a)** follows.

Now let $f \in A(E,1)$ satisfy $f'(\infty) = \gamma(E)$. Then

$$\int_{\partial U} f(\zeta)\psi(\zeta)\, d\zeta = \gamma(E) = \int_{\partial U} |\psi(\zeta)|\, ds.$$

This equality implies that

(i) $f(\zeta)\psi(\zeta)d\zeta \geq 0$ on ∂U;

and, because ψ cannot vanish on a set of positive measure,

(ii) $|f(\zeta)| = 1$ a.e. on ∂U.

Condition **(i)** and the analyticity of each Γ_j imply that $f(z)\psi(z)$ is analytic across ∂U. For, if $\zeta(w)$ is analytic on $\{r < |w| < 1/r\}$, with $\zeta(\{|w| = 1\}) = \Gamma_j$, then (for r near 1) $g(w) = f(\zeta)\psi(\zeta)$ is in H^1 of the annulus $A = \{1 < |w| < 1/r\}$. Also

$g(w)\zeta'(w) \geq 0$ on $\{|w| = 1\}$. Schwarz reflection and Morera's theorem can now be used to extend $f(w)\zeta'(w)$ to the larger annulus $\{r < |w| < 1/r\}$ so that $f(\zeta)\psi(\zeta)$ is analytic on ∂U. Now the argument principle and condition **(i)** above imply that

(iii) $f(z)\psi(z)$ has k zeros on U.

For each $\zeta \in \Gamma$, take a neighborhood V of ζ such that $V \cap U$ is simply connected, $V \cap \Gamma$ is an arc, and $f(z)\psi(z)$ has no zeros on $V \cap U$. After applying a conformal map $z = z(w)$, we have nonvanishing functions $\widetilde{f}(w)$ and $\widetilde{\psi}(w)$ on the unit disc D and an arc J in ∂D (corresponding to $V \cap \Gamma$) on which $\widetilde{f}(w)\widetilde{\psi}(w)$ is continuous. Write $\widetilde{f} = S_1 F_1$ and $\widetilde{\psi} = S_2 F_2$, where each S_j is a singular inner function and each F_j is outer (see [HOF] for these ideas). Then, because $\widetilde{f}\widetilde{\psi}$ is analytic across J, the same holds for $S_1 \cdot S_2$ and hence for S_1. Since

$$F_1(w) = \exp\left(\frac{1}{2\pi}\int_{\partial U} \frac{e^{i\theta} + w}{e^{i\theta} - w} \log|F_1(e^{i\theta})|\, d\theta\right),$$

and $|F_1| = 1$ on J by **(ii)**, we find that F_1 is analytic across J also. Therefore $\widetilde{f}$ is analytic across J and $|\widetilde{f}| = 1$ on J. Returning to V, we see that we have proved **(b)** and **(c)**.

Finally, **(b)** and **(c)** tell us that f has at least k zeros on U while **(iii)** ensures that there can be no more than k zeros.

That ends the proof. □

COROLLARY 8.17 *When ∂U consists of finitely many pairwise disjoint, analytic Jordan curves then the Garabedian function ψ is analytic across ∂U and $\log \psi$ is single valued on U.*

Prelude: This result tells us a way to find extremal functions for analytic capacity.

THEOREM 8.18 *For any compact set $E \subseteq \mathbb{C}$ there is a unique function $f \in A(E, 1)$ such that $f'(\infty) = \gamma(E)$.*

Proof: Let E_j be compact sets decreasing to E so that ∂E_j consists of finitely many pairwise disjoint, analytic curves. Let

$$f_j(z) = \frac{\gamma(E_j)}{z} + \cdots$$

be any Ahlfors function for E_j and let

$$\psi_j(z) = \frac{1}{2\pi i} + \cdots$$

be a Garabedian function for E_j. Then, for $g \in A(E_j, 1)$ and for $z \in U_j \equiv \mathbb{C} \setminus E_j$, we have (by Cauchy's theorem)

$$\int_{\partial U_j} \frac{g(z) - g(\zeta)}{z - \zeta} \psi_j(\zeta)\, d\zeta = g(z).$$

Therefore

$$g(z) = \left(\oint \frac{\psi_j(\zeta)}{\zeta - z}\, d\zeta - 1 \right)^{-1} \cdot \oint \frac{g(\zeta)\psi_j(\zeta)}{\zeta - z}\, d\zeta.$$

Choose k_j so that the sequence $\{f_{k_j}\}$ converges pointwise to a function $f \in A(E, 1)$. The sequence $\{\psi_{k_j}(\zeta)\, d\zeta\}$ then converges in the weak-$*$ topology to a measure σ on E. Also the sequence

$$\tau_{k_j} = \{f_{k_j}(\zeta)\psi_{k_j}(\zeta)\, d\zeta\} = \{|\psi_k(\zeta)|\, |d\zeta|\}$$

converges in the weak-$*$ topology to a positive measure τ on E. Thus

$$\begin{aligned} f(z) &= \lim_{j\to\infty} \left(\oint \frac{\psi_{k_j}(\zeta)\, d\zeta}{\zeta - z} - 1 \right)^{-1} \cdot \oint \frac{f_{k_j}(\zeta)\psi_{k_j}(\zeta)}{\zeta - z}\, d\zeta \\ &= \left(\int_E \frac{d\sigma(\zeta)}{\zeta - z} - 1 \right)^{-1} \cdot \int_E \frac{d\tau(\zeta)}{\zeta - z}. \end{aligned}$$

Let g be any extremal function for E with $g \in A(E, 1)$, $g'(\infty) = \gamma(E)$. Then

$$g(z) = \left(\int_E \frac{d\sigma(\zeta)}{\zeta - z} - 1 \right)^{-1} \cdot \lim_{j\to\infty} \oint \frac{g(\zeta)\psi_{k_j}(\zeta)}{\zeta - z}\, d\zeta.$$

We assert that the measures $\lambda_j = g(\zeta)\psi_{k_j}(\zeta)d\zeta$ converge weak-$*$ to ψ. That clearly implies that $g = f$. Since

$$|g(\zeta)\psi_{k_j}(\zeta)d\zeta| \leq |\psi_{k_j}(\zeta)||d\zeta|,$$

we have $|\lambda_j| \leq \tau_{k_j}$. On the other hand

$$\int d\lambda_j = \int g(\zeta)\psi_{k_j}(\zeta)\, d\zeta = \gamma(E) = \lim_{j\to\infty} \int d\tau_{k_j} = \int d\tau.$$

Thus if λ is a weak-$*$ limit of $\{\lambda_j\}$, then $|\lambda| \leq \tau$ and $\int d\lambda = \int d\tau$. Therefore $\lambda = \tau$. That means that λ_j converges to τ. □

Prelude: As previously noted, additivity properties for analytic capacity are hard to come by. So this is a fundamental result.

THEOREM 8.19 (POMMERENKE) *Let E_1, E_2, ..., E_k be pairwise disjoint, compact sets in $\mathbb{C}$ and set $K = E_1 + E_2 + \cdots + E_k$ be their Minkowski sum.*[2] *Then*

$$\gamma\left(\bigcup_{j=1}^{k} E_j\right) \leq \gamma(K).$$

Proof: Let U be the complement of $E \equiv \cup_{j=1}^{k} E_j$. Let $f \in A(E,1)$ satisfy $f'(\infty) = \gamma(E)$. Conformally mapping U to a domain with real analytic boundary (for instance, a large disc with smaller discs removed), we see by results of Theorem 8.16 above that

(i) f is a k-to-1 covering map of D.

(ii) f is proper (in the sense of topology). That is to say, if $U \ni z_j \to z \in E$, then $|f(z_j)| \to 1$.

Choose a number ρ, $0 < \rho < 1$, such that $L \equiv f^{-1}(\{|w| = \rho\})$ consists of k analytic curves and such that there are inverses φ_1, φ_2, ..., φ_k of f mapping $\{\rho < |w| < 1\}$ onto the k annular components of $\{z : \rho < |f(z)| < 1\}$. Now Cauchy's theorem tells us that

$$\frac{1}{2\pi i}\oint_L z(f(z))^n f'(z)\, dz = -\gamma(E)\delta_{n,0}$$

for $n \geq 0$. (Here $\delta_{n,0}$ is the Kronecker delta.) Thus

$$\sum_{j=1}^{k} \frac{1}{2\pi i}\oint_{|w|=\rho} \varphi_j(w) w^n\, dw = \gamma(E) \cdot \delta_{n,0}.$$

Also

$$\Phi(w) \equiv \sum_{j=1}^{k} \varphi_j(w)$$

extends to be meromorphic on $\{|w| < 1\}$ with a simple pole at 0:

$$\Phi(w) = \frac{\gamma(E)}{w} + a_0 + a_1 w + \cdots.$$

Let $T = \{\lim \Phi(w_k) : |w_k| \to 1\}$. Then T is a compact subset of K. Let U be the component of $D \setminus \Phi^{-1}(T)$ containing the origin.

[2]The plus signs denote the Minkowski sum of sets. Here the Minkowski sum $A + B$ of two sets A and B is defined to be $A + B = \{a + b : a \in A, b \in B\}$.

Then Φ is a proper mapping of U onto $\Phi(U)$ and Φ has a simple pole at 0. Thus, by the argument principle, Φ is univalent. Also $\partial\Phi(U) = T \subset K$ and Φ^{-1} is in $A(T,1)$ with derivative $\gamma(E)$ at ∞. Thus $\gamma(E) \leq \gamma(T) \leq \gamma(K)$. □

Prelude: Now we have a very concrete result about analytic capacity for subsets of the real line.

THEOREM 8.20 *Let E be a subset of the real axis $\mathbb{R}$. Then*

$$\gamma(E) = \frac{\ell(E)}{4},$$

where $\ell(E)$ is the linear measure of E.

Proof: We already know this result for E an interval. Now assume that E is the pairwise disjoint union of closed intervals $\{E_j\}_{j=1}^k$. Then

$$K \equiv E_1 + E_2 + \cdots + E_k$$

is a closed interval of length $\ell(E)$. By Pommerenke's theorem,

$$\gamma(E) \leq \gamma(K) = \frac{\ell(E)}{4}.$$

Approximating from above yields the same inequality for all compact sets E; approximating from below yields the result for all E.

Conversely, let E be compact, and write

$$f(z) = \frac{1}{2}\int_E \frac{dt}{t-z}.$$

Then $f(\infty) = 0$, $f'(\infty) = -\ell(E)/2$, and

$$|\text{Im}\, f(z)| = \frac{|y|}{2}\int_E \frac{dt}{(t-x)^2+y^2}, \qquad z = x+iy.$$

As a result,

$$|\text{Im}\, f(z)| \leq \frac{1}{2}\int_{-\infty}^{\infty} \frac{du}{1+u^2} = \frac{\pi}{2}.$$

Thus $\text{Re}\, e^{f(z)} > 0$ and

$$F(z) = \frac{1-e^{f(z)}}{1+e^{f(z)}} = \frac{\ell(E)}{4z} + \frac{a_2}{z^2} + \cdots$$

satisfies $|F(z)| \leq 1$. Therefore $\gamma(E) \geq \ell(E)/4$, and that is our result. □

Then, Φ is a proper mapping of U onto $\Omega(T)$ and Φ has a simple pole at 0. Thus, by the argument principle, Φ is univalent. Also $\partial\Phi(U) = \ldots \subset T$ and $\ldots$ is in $A(T, 1)$ with derivative $\gamma(E)$ at ∞. Thus $\gamma(E) \leq \gamma(T) \leq \gamma(E)$. □

Prelude: Now we have a very concrete result about analytic capacity for subsets of the real line.

THEOREM 8.20 *Let E be a subset of the real axis $\mathbb{R}$. Then*

$$\gamma(E) = \frac{\ell(E)}{4},$$

where $\ell(E)$ is the linear measure of E.

Proof. We already know this result for E an interval. Now assume that E is the union of disjoint union of closed intervals $\{E_j\}_{j=1}^{n}$. Then

$$E' = E_1 + E_2 + \cdots + E_n$$

is a closed interval of length $\ell(E)$. By Pommerenke's theorem,

$$\gamma(E) \geq \gamma(E') = \frac{\ell(E)}{4}.$$

Approximating from above yields the same inequality for all compact sets E; approximating from below yields the result for all E.

Conversely, let E be compact and write

$$f(z) = \frac{1}{2}\int_E \frac{dt}{z-t}.$$

Then $f(\infty) = 0$, $f'(\infty) = \ell(E)/2$, and

$$\operatorname{Im} f(z) = \frac{-y}{2}\int_E \frac{dt}{(x-t)^2+y^2}, \quad z = x+iy.$$

As a result,

$$|\operatorname{Im} f(z)| \leq \frac{|y|}{2}\int_{-\infty}^{\infty}\frac{dt}{t^2+y^2} = \frac{\pi}{2}.$$

Thus $\operatorname{Re}(\ldots) > 0$ and

$$F(z) = \frac{1-e^{f(z)}}{1+e^{f(z)}} = \frac{\ell(E)}{4z} + \cdots$$

satisfies $|F(z)| \leq 1$. Therefore $\gamma(E) \geq \ell(E)/4$, [illegible] result. □

Chapter 9

Invariant Geometry

Prologue: One of the profound ideas of twentieth century complex analysis is the Bergman theory. Inspired by early studies of E. Schmidt and others of L^2 of the unit interval, Bergman introduced his now celebrated space of square integrable holomorphic functions on a given domain U. He showed that this is a Hilbert space, and he proved that there is a reproducing kernel for this space. This kernel is now known as the Bergman kernel.

The Bergman kernel is important both for its reproducing properties and its invariance properties. Perhaps even more important is that the kernel can be used to produce an invariant metric known as the Bergman metric. The Bergman metric, historically speaking, was the first example of a Kähler metric. Kähler geometry is one of the most active research areas of modern mathematics.

In his lifelong studies, Bergman related his kernel and geometry to mapping theory, to elliptic partial differential equations, to differential geometry, to extremal problems, and to many other different parts of mathematics.

The Bergman theory was given great prominence by C. Fefferman's Fields-Medal-winning work on the boundary smoothness of biholomorphic mappings of several complex variables. Since then, S. Bell, E. Ligocka, T. Ohsawa, R. E. Greene, S. G. Krantz, and many others have contributed notable developments to the Bergman theory.

This chapter will be your first instroduction to the profound ideas of Stefan Bergman.

9.1 Conformality and Invariance

Capsule: As predicted earlier, we "discover" here the Poincaré metric by means of some simple-minded calculations.

Conformal mappings are characterized by the fact that they infinitesimally **(i)** preserve angles, and **(ii)** preserve length (up to a scalar factor). It is worthwhile to picture the matter in the following manner: Let f be holomorphic on the open set $U \subseteq \mathbb{C}$. Fix a point $P \in U$. Write $f = u + iv$ as usual. Thus we may write the mapping f as $(x, y) \mapsto (u, v)$. Then the (real) Jacobian matrix of the mapping is

$$J(P) = \begin{pmatrix} u_x(P) & u_y(P) \\ v_x(P) & v_y(P) \end{pmatrix},$$

where subscripts denote derivatives. We may use the Cauchy-Riemann equations to rewrite this matrix as

$$J(P) = \begin{pmatrix} u_x(P) & u_y(P) \\ -u_y(P) & u_x(P) \end{pmatrix}.$$

Factoring out a numerical coefficient, we finally write this two-dimensional derivative as

$$\begin{aligned} J(P) &= \sqrt{u_x(P)^2 + u_y(P)^2} \\ &\quad \cdot \begin{pmatrix} \frac{u_x(P)}{\sqrt{u_x(P)^2+u_y(P)^2}} & \frac{u_y(P)}{\sqrt{u_x(P)^2+u_y(P)^2}} \\ \frac{-u_y(P)}{\sqrt{u_x(P)^2+u_y(P)^2}} & \frac{u_x(P)}{\sqrt{u_x(P)^2+u_y(P)^2}} \end{pmatrix} \\ &\equiv h(P) \cdot \mathcal{J}(P). \end{aligned}$$

The matrix $\mathcal{J}(P)$ is a special orthogonal matrix (i.e., its rows form an orthonormal basis of $\mathbb{R}^2$, and it is oriented positively). Thus we see that the derivative of our mapping is a rotation $\mathcal{J}(P)$ (which preserves angles) followed by a positive "stretching factor," $h(P)$ (which also preserves angles).

It would be incorrect to infer from these considerations that a conformal map therefore preserves Euclidean angles, or Euclidean length, in any global sense. Such a mapping would perforce be linear and, in fact, special orthogonal; and thus complex function theory

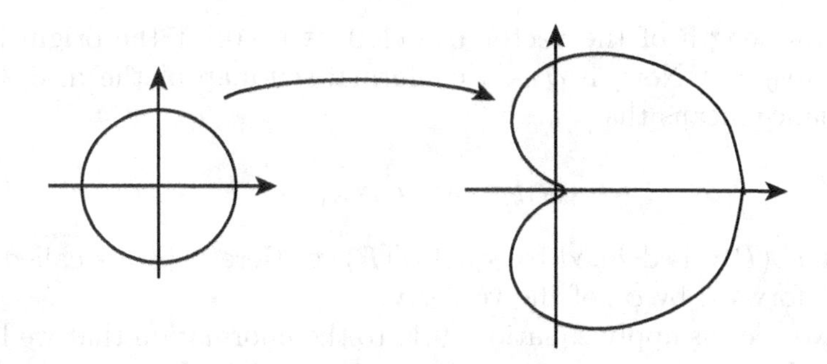

Figure 9.1: The failure of Euclidean isometry under conformal mappings.

would be reduced to a triviality. The mapping of the disc $D(0,2)$ given by

$$\phi : z \mapsto (z+4)^4,$$

illustrated in Figure 9.1, exhibits rather dramatically the failure of Euclidean isometry. The mapping is one-to-one, onto its image, yet $\phi(0) = 256$, $\phi(1) = 625$, and hence

$$1 - 0 = 1 \neq 369 = \phi(1) - \phi(0).$$

And yet it is obviously desirable to have a notion of distance that *is preserved* under holomorphic mappings. Part of Klein's *Erlangen program* is to understand geometric objects according to the groups that act on them. Such an understanding is facilitated when the group is actually a group of isometries.

It was H. Poincaré who first found a way to carry out this idea when the domain in question is the unit disc D. To understand his thinking, let us recall that the collection of all conformal maps of the unit disc $D = D(0,1)$ can be described explicitly: it consists of

(i) all rotations $\rho_\lambda : z \mapsto e^{i\lambda} \cdot z$, $0 \le \lambda < 2\pi$;

(ii) all Möbius transformations $\varphi_a : z \mapsto [z-a]/[1-\overline{a}z]$, $a \in \mathbb{C}$, $|a| < 1$;

(iii) all compositions of mappings of type **(i)** and **(ii)**.

We will use Riemann's paradigm for a metric, that is, we shall specify the length of a (tangent) vector at each point of the disc D. If the point is P and the vector is $\mathbf{v}$, then let us denote this length by $|\mathbf{v}|_P$.

Our goal now is to "discover" the Poincaré metric by way of a sequence of calculations. We must begin somewhere, so let us declare that the length of the vector $\mathbf{e} \equiv (1,0) \equiv 1+0i$ at the origin is 1. Thus $|\mathbf{e}|_0 = 1$. Now, if ϕ is a conformal self-map of the disc, then invariance means that

$$|\mathbf{v}|_P = |\phi_*(P)\mathbf{v}|_{\phi(P)}, \tag{9.1}$$

where $\phi_*(P)\mathbf{v}$ is defined to equal $\phi'(P) \cdot \mathbf{v}$. Here $\phi_*(P)$ is called the "push-forward by ϕ" of the vector $\mathbf{v}$.

Now let us apply equation (9.1) to the information that we have about the length of the unit vector $\mathbf{e}$ at the origin. Let $\phi(z) = e^{i\lambda} \cdot z$. Then we see that

$$1 = |\mathbf{e}|_0 = |\phi_*(0)\mathbf{e}|_{\phi(0)} = |e^{i\lambda} \cdot \mathbf{e}|_0 = |e^{i\lambda}|_0. \tag{9.2}$$

We conclude that the length of the vector $e^{i\lambda}$ at the origin is 1. So all Euclidean unit vectors based at the origin have length 1 in our new invariant metric.

Now let $\psi(z)$ be the Möbius transformation

$$\psi(z) = \frac{z+a}{1+\overline{a}z},$$

with a a complex number of modulus less than 1. Let $\mathbf{v} = e^{i\lambda}$ be a unit vector at the origin. Then $\psi'(0) = 1 - |a|^2$ and equation (9.1) tells us that

$$1 = |\mathbf{v}|_0 = |\psi_*(0)\mathbf{v}|_{\psi(0)} = |(1-|a|^2) \cdot \mathbf{v}|_a.$$

We conclude that

$$|\mathbf{v}|_a = \frac{1}{1-|a|^2}.$$

Of course our new metric will respect scalar multiplication, so we may apply the preceding calculation to vectors of any length. If we let $\|\mathbf{v}\|$ denote the *Euclidean length* of a vector $\mathbf{v}$, then we may summarize all our calculations as follows.

> Let P be a point of the unit disc D, and let $\mathbf{v}$ be any vector based at that point. Then
>
> $$|\mathbf{v}|_P = \frac{\|\mathbf{v}\|}{1-|P|^2}.$$
>
> We call this new metric the *Poincaré metric.*

We see, in particular that, as P gets ever nearer to the boundary of D, the new metric length of $\mathbf{v}$ becomes greater and greater.

It is well worth spending some time interpreting this new metric and its implications for complex analysis. Let $\gamma : [0,1] \to D$ be a continuously differentiable curve. We define the *length* of γ in the Poincaré metric to be

$$\ell(\gamma) = \int_0^1 |\gamma'(t)|_{\gamma(t)}\, dt.$$

Observe that $\gamma'(t)$ is a vector located at $\gamma(t)$, so the definition makes good sense. We can consider curves parametrized over *any* interval, and the definition of length is independent (by the change of variables formula of calculus) of the choice of parametrization. We define the length of a *piecewise continuously differentiable* curve to be the sum of the lengths of its continuously differentiable pieces.

Example 9.3 Let $\epsilon > 0$. Consider the curve $\gamma(t) = (1-\epsilon)t$, $0 \le t \le 1$. Then, according to the definition,

$$\begin{aligned}\ell(\gamma) = \int_0^1 |\gamma'(t)|_{\gamma(t)}\, dt &= \int_0^1 \frac{(1-\epsilon)}{1-|\gamma(t)|^2}\, dt \\ &= \int_0^1 \frac{(1-\epsilon)}{1-[(1-\epsilon)t]^2}\, dt = \frac{1}{2}\log\left(\frac{2-\epsilon}{\epsilon}\right).\end{aligned}$$

We see immediately that, as $\epsilon \to 0^+$, the expression $\ell(\gamma)$ tends to $+\infty$. Thus, in some sense, the distance from 0 to the boundary (at least along the given linear path) is infinite.

We define the *Poincaré distance* $d(P,Q)$ between two points $P, Q \in D$ to be the infimum of the lengths of all piecewise continuously differentiable curves connecting P to Q.

PROPOSITION 9.4 *Let $P \in D$. Then the Poincaré distance of 0 to P is equal to*

$$d(0,P) = \frac{1}{2} \cdot \log \frac{1+|P|}{1-|P|}.$$

Proof: By rotational invariance, we may as well suppose that P is real and positive, so $P = (1-\epsilon) + i0 \equiv (1-\epsilon, 0)$. It is an exercise for the reader to see that there is no loss of generality to consider only curves of the form $\gamma(t) = (t, g(t))$, $0 \le t \le 1 - \epsilon$. Then

$$\begin{aligned}\ell(g) &= \int_0^{1-\epsilon} \frac{\|\gamma'(t)\|}{1 - |\gamma(t)|^2}\,dt = \int_0^{1-\epsilon} \frac{\sqrt{1^2 + |g'(t)|^2}}{1 - t^2 - |g^2(t)|}\,dt \\ &\ge \int_0^{1-\epsilon} \frac{1}{1-t^2}\,dt = \frac{1}{2} \cdot \log\left(\frac{2-\epsilon}{\epsilon}\right) \\ &= \frac{1}{2} \cdot \log\left(\frac{1+|P|}{1-|P|}\right).\end{aligned}$$

Thus we see explicitly that $\mu(t) = (t, 0)$ is the shortest curve from 0 to P, and the distance is as we claimed. □

Combining the result of this proposition with the preceding example, we see that any curve that starts at the origin and runs out to the boundary of the disc will have infinite length. Thus the boundary is infinitely far away.

Exercise for the Reader: Prove that if P is any point of the disc and $r > 0$, then the metric disc

$$\beta(P, r) \equiv \{z \in D : d(z, P) < r\}$$

is actually a Euclidean disc. What are its Euclidean center and radius (expressed in terms of r and P)? Show that the disc $\beta(P, r)$ is relatively compact (i.e., has compact closure) in the disc. It follows that any Cauchy sequence *in the Poincaré metric* has a limit point *in the disc.* Thus the Poincaré metric turns D into a complete metric space. □

9.2 Bergman's Construction

Capsule: Here we give Bergman's construction of the Bergman space and also of the Bergman metric. Along the way, we prove a number of very special properties of these constructs.

Stefan Bergman created a device for equipping virtually *any* planar domain with an invariant metric that is analogous to the Poincaré metric on the disc.[1] Some tracts call this new metric the *Poincaré-Bergman metric*, though it is more commonly called just

[1]In Section 1.7, we treated the uniformization theorem of Koebe. It gives a means for transferring the Poincaré metric from the disc to virtually any planar domain.

the *Bergman metric.* In order to construct the Bergman metric we must first construct the Bergman kernel. For that we need just a little Hilbert space theory (see [RUD2], for example).

A *domain* in $\mathbb{C}$ is a connected open set. Fix a domain $U \subseteq \mathbb{C}$, and define

$$A^2(U) = \left\{ f \text{ holomorphic on } U : \int_U |f(z)|^2 \, dA(z) < \infty \right\} \subseteq L^2(U).$$

Here dA is ordinary two-dimensional area measure. Of course $A^2(U)$ is a complex linear space, called the *Bergman space.* The norm on $A^2(U)$ is given by

$$\|f\|_{A^2(U)} = \left[\int_U |f(z)|^2 \, dA(z) \right]^{1/2}.$$

We define an inner product on $A^2(U)$ by

$$\langle f, g \rangle = \int_U f(z)\overline{g(z)} \, dA(z).$$

The next technical lemma will be the key to our analysis of the space A^2.

LEMMA 9.5 *Let $K \subseteq U$ be compact. There is a constant $C_K > 0$, depending on K, such that*

$$\sup_{z \in K} |f(z)| \leq C_K \|f\|_{A^2(U)}, \quad \text{all } f \in A^2(U).$$

Proof: Since K is compact, there is an $r(K) = r > 0$ so that, for any $z \in K, D(z,r) \subseteq U$. Therefore, for each $z \in K$ and $f \in A^2(U)$, we may use the mean value property of harmonic functions to see that

$$\begin{aligned} |f(z)| &= \frac{1}{A(B(z,r))} \left| \int_{B(z,r)} f(t) \, dA(t) \right| \\ &\leq \frac{1}{A(B(z,r))} \cdot \int_{\mathbb{C}} \chi_{B(z,r)}(t) \cdot |f(t)| \, dA(t), \end{aligned}$$

where

$$\chi_{B(z,r)}(t) = \begin{cases} 1 & \text{if} \quad t \in B(z,r) \\ 0 & \text{if} \quad t \notin B(z,r). \end{cases}$$

We apply the Cauchy-Schwarz inequality from integration theory (see [RUD2]) to the last expression to find that it is less than or equal to

$$\begin{aligned}\frac{1}{A(B(z,r))}\cdot \int_{\mathbb{C}} |\chi_{B(z,r)}(t)|^2\, dA^{1/2} \cdot \int_{\mathbb{C}} |f(t)|^2\, dA^{1/2} \\ &= (A(B(z,r)))^{-1/2}\|f\|_{A^2(B(z,r))} \\ &= \frac{1}{\sqrt{\pi} r}\|f\|_{A^2(U)} \\ &\equiv C_K\|f\|_{A^2(U)}.\end{aligned}$$

□

PROPOSITION 9.6 *The space $A^2(U)$ is complete.*

Proof: Let $\{f_j\}$ be a Cauchy sequence in A^2. Then the sequence is Cauchy in $L^2(U)$, and the completeness of L^2 (see [RUD2]) then tells us that there is a limit function f. So $f_j \to f$ in the L^2 topology. Now the lemma tells us that the convergence is taking place uniformly on compact sets. So $f \in A^2(U)$. That completes the argument. □

COROLLARY 9.7 *The space $A^2(U)$ is a Hilbert space.*

With a little extra effort (see [GRK1], [RUD3]), it can be shown that $A^2(U)$ is in fact a *separable* Hilbert space.

Now we need to find a method for representing certain linear functionals. The key fact is this:

LEMMA 9.8 *For each fixed $z \in U$, the functional*

$$\Phi_z : f \mapsto f(z), \quad f \in A^2(U)$$

is a continuous linear functional on $A^2(U)$.

Proof: This is immediate from Lemma 9.5 if we take K to be the singleton $\{z\}$. □

We may now apply the Riesz representation theorem (see [RUD2]) to see that there is an element $k_z \in A^2(U)$ such that the linear functional Φ_z is represented by inner product with k_z. If $f \in A^2(U)$ then, for all $z \in U$, we have

$$f(z) = \Phi_z(f) = \langle f, k_z \rangle. \tag{9.9}$$

Definition 9.10 The *Bergman kernel* is the function $K(z,\zeta) = \overline{k_z(\zeta)}$, $z, \zeta \in U$. It has the reproducing property

$$f(z) = \int K(z,\zeta) f(\zeta)\, dA(\zeta), \quad \forall f \in A^2(U). \tag{9.10.1}$$

We sometimes denote the Bergman kernel for U by $K_U(z,\zeta)$.

Notice that (9.10.1) is just a restatement of (9.9).

PROPOSITION 9.11 *The Bergman kernel $K(z,\zeta)$ is conjugate symmetric: $K(z,\zeta) = \overline{K(\zeta,z)}$.*

Proof: By its very definition, $\overline{K(\zeta,\cdot)} \in A^2(U)$ for each fixed ζ. Therefore the reproducing property of the Bergman kernel gives

$$\int_U K(z,t)\overline{K(\zeta,t)}\, dA(t) = \overline{K(\zeta,z)}.$$

On the other hand,

$$\begin{aligned}\int_U K(z,t)\overline{K(\zeta,t)}\, dA(t) &= \overline{\int K(\zeta,t)\overline{K(z,t)}\, dA(t)} \\ &= \overline{\overline{K(z,\zeta)}} = K(z,\zeta). \qquad \square\end{aligned}$$

PROPOSITION 9.12 *The Bergman kernel is uniquely determined by the properties that it is an element of $A^2(U)$ in z, is conjugate symmetric, and reproduces $A^2(U)$.*

Proof: Let $\widetilde{K}(z,\zeta)$ be another such kernel. Then

$$\begin{aligned}K(z,\zeta) = \overline{K(\zeta,z)} &= \int \widetilde{K}(z,t)\overline{K(\zeta,t)}\, dA(t) \\ &= \overline{\int K(\zeta,t)\overline{\widetilde{K}(z,t)}\, dA(t)} = \overline{\overline{\widetilde{K}(z,\zeta)}} = \widetilde{K}(z,\zeta). \qquad \square\end{aligned}$$

Since $A^2(U)$ is a separable Hilbert space, there is a complete orthonormal basis $\{\phi_j\}_{j=1}^{\infty}$ for $A^2(U)$.

PROPOSITION 9.13 *Let E be a compact subset of U. Then the series*

$$\sum_{j=1}^{\infty} \phi_j(z)\overline{\phi_j(\zeta)}$$

sums uniformly on $E \times E$ to the Bergman kernel $K(z,\zeta)$.

Proof: By the Riesz-Fisher and Riesz representation theorems, we obtain

$$\sup_{z\in E}\left(\sum_{j=1}^{\infty}|\phi_j(z)|^2\right)^{1/2} = \sup_{z\in E}\left\|\{\phi_j(z)\}_{j=1}^{\infty}\right\|_{\ell^2}$$

$$= \sup_{\substack{\|\{a_j\}\|_{\ell^2}=1 \\ z\in E}}\left|\sum_{j=1}^{\infty}a_j\phi_j(z)\right|$$

$$= \sup_{\substack{\|f\|_{A^2}=1 \\ z\in E}}|f(z)| \le C_E. \tag{9.13.1}$$

In the last inequality we have used Lemma 9.5. Therefore

$$\sum_{j=1}^{\infty}\left|\phi_j(z)\overline{\phi_j(\zeta)}\right| \le \left(\sum_{j=1}^{\infty}|\phi_j(z)|^2\right)^{1/2}\left(\sum_{j=1}^{\infty}|\phi_j(\zeta)|^2\right)^{1/2}$$

and the convergence is uniform over $z, \zeta \in E$. For fixed $z \in U$, our preceding calculation shows that $\{\phi_j(z)\}_{j=1}^{\infty} \in \ell^2$. Hence we have that $\sum \phi_j(z)\overline{\phi_j(\zeta)} \in \overline{A^2(U)}$ as a function of ζ. Let the sum of the series be denoted by $\widetilde{K}(z,\zeta)$. Notice that $\widetilde{K}$ is conjugate symmetric by its very definition. Also, for $f \in A^2(U)$, we have

$$\int \widetilde{K}(\cdot,\zeta)f(\zeta)\,dA(\zeta) = \sum \widehat{f}(j)\phi_j(\cdot) = f(\cdot),$$

where convergence is in the Hilbert space topology. (Here $\widehat{f}(j)$ is the jth Fourier coefficient of f with respect to the basis $\{\phi_j\}$.) But Hilbert space convergence dominates pointwise convergence (Lemma 9.5), so

$$f(z) = \int \widetilde{K}(z,\zeta)f(\zeta)\,dA(\zeta), \quad \text{all } f \in A^2(U).$$

Therefore, by Proposition 9.12, $\widetilde{K}$ is the Bergman kernel. □

PROPOSITION 9.14 *If U is a bounded domain in $\mathbb{C}$, then the mapping*

$$P : f \mapsto \int_U K_U(\cdot,\zeta)f(\zeta)\,dA(\zeta)$$

is the Hilbert space orthogonal projection of $L^2(U, dA)$ onto $A^2(U)$.

Proof: Notice that P is idempotent and self-adjoint and that $A^2(U)$ is precisely the set of elements of L^2 that are fixed by P. □

LEMMA 9.15 *Let $\phi : U_1 \to U_2$ be a one-to-one, onto, conformal mapping of domains. Then*

$$\int_{U_1} |\phi'(z)|^2\, dx\, dy = \int_{U_2} dx\, dy.$$

Proof: Write $\phi = u + iv$. Then the Jacobian determinant of ϕ, thought of as a real mapping from $\mathbb{R}^2$ to $\mathbb{R}^2$, is

$$J = \det \begin{pmatrix} u_x & u_y \\ v_x & v_y \end{pmatrix},$$

which, by the Cauchy-Riemann equations, equals

$$\det \begin{pmatrix} u_x & -v_x \\ v_x & u_x \end{pmatrix} = u_x^2 + v_x^2 = |\phi_x|^2 = |\phi'|^2.$$

The last equality is obtained by another application of the Cauchy-Riemann equations.

As a result,

$$\iint_{\Omega_2} 1\, dA = \iint_{\Omega_1} J\, dA = \iint_{\Omega_1} |\phi'(z)|^2\, dx\, dy. \qquad \square$$

This last result is called the *Lusin area integral formula.*

PROPOSITION 9.16 *Let U_1, U_2 be domains in $\mathbb{C}$. Let $f : U_1 \to U_2$ be conformal. Then*

$$f'(z) K_{U_2}(f(z), f(\zeta)) \overline{f'(\zeta)} = K_{U_1}(z, \zeta).$$

Proof: Let $\phi \in A^2(U_1)$. Then, by change of variable,

$$\int_{U_1} \left[f'(z) K_{U_2}(f(z), f(\zeta)) \overline{f'(\zeta)} \right] \phi(\zeta)\, dA(\zeta)$$

$$= \int_{U_2} f'(z) K_{U_2}(f(z), \widetilde{\zeta}) \overline{f'(f^{-1}(\widetilde{\zeta}))} \phi(f^{-1}(\widetilde{\zeta})) \cdot \frac{1}{|f'(f^{-1}(\widetilde{\zeta}))|^2}\, dA(\widetilde{\zeta}).$$

In the last equality, we have used Lemma 9.15. The righthand side simplifies to

$$f'(z) \int_{U_2} K_{U_2}(f(z), \widetilde{\zeta}) \left\{ \frac{1}{f'(f^{-1}(\widetilde{\zeta}))} \phi\left(f^{-1}(\widetilde{\zeta})\right) \right\} dA(\widetilde{\zeta}).$$

By change of variables, the expression in braces { } is an element of $A^2(U_2)$. So the reproducing property of K_{U_2} applies and the last line equals

$$f'(z) \cdot \frac{1}{f'(z)} \cdot \phi(z) = \phi(z).$$

By the uniqueness of the Bergman kernel, the proposition follows. □

PROPOSITION 9.17 *For $z \in U \subset\subset \mathbb{C}$ it holds that $K_U(z,z) > 0$.*

Proof: If $\{\phi_j\}$ is a complete orthonormal basis for $A^2(U)$ then

$$K_\omega(z,z) = \sum_{j=1}^{\infty} |\phi_j(z)|^2 \geq 0.$$

If $K(z,z) = 0$ for some z, then $\phi_j(z) = 0$ for all j; hence $f(z) = 0$ for every $f \in A^2(U)$. This is absurd. □

As an exercise, the reader may wish to verify that the last proposition is true with only the simple assumption that the area is finite.

Definition 9.18 For any bounded $U \subseteq \mathbb{C}$, we define a Hermitian metric on U by

$$g(z) = \frac{\partial^2}{\partial z \partial \overline{z}} \log K(z,z), \quad z \in U.$$

This means that the length of a tangent vector ξ at a point $z \in U$ is given by

$$|\xi|_z = g(z)\|\xi\|.$$

The metric that we have defined is called the *Bergman metric.* Sometimes, for clarity, we denote the metric by g^U.

We note that the length of a curve, and the distance $d_\omega(z,w)$ between two points z, w, is now defined just as for the Poincaré metric on the disc.

PROPOSITION 9.19 *Let $U_1, U_2 \subseteq \mathbb{C}$ be domains and let $f : U_1 \to U_2$ be a conformal mapping. Then f induces an isometry of Bergman metrics:*

$$|\xi|_z = |f'(z) \cdot \xi|_{f(z)} \tag{9.19.1}$$

for all $z \in U_1, \xi \in \mathbb{C}$. Equivalently, f induces an isometry of Bergman distances in the sense that

$$d_{U_2}(f(P), f(Q)) = d_{U_1}(P,Q).$$

Proof: This is a formal exercise, but we include it for completeness.

From the definitions, it suffices to check that

$$|g^{U_2}(f(z))f'(z)w| = |g^{U_1}(z)w|$$

for all $z \in U, w \in \mathbb{C}$. But, by Proposition 9.16,

$$\begin{aligned} g^{U_1}(z) &= \frac{\partial^2}{\partial z \partial \bar{z}} \log K_{U_1}(z,z) \\ &= \frac{\partial^2}{\partial z \partial \bar{z}} \log \left\{|f'(z)|^2 K_{U_2}(f(z), f(z))\right\} \\ &= \frac{\partial^2}{\partial z \partial \bar{z}} \log K_{U_2}(f(z), f(z)) \end{aligned} \tag{9.19.2}$$

since, locally,

$$\log|f'(z)|^2 = \log\left(f'(z)\right) + \log\left(\overline{f'(z)}\right) + C.$$

But line (9.19.2) is nothing other than

$$|g^{U_2}(f(z))| \frac{\partial f(z)}{\partial z} \frac{\partial \overline{f(z)}}{\partial \bar{z}} = |g^{U_2}(f(z))| \left|\frac{\partial f(z)}{\partial z}\right|^2,$$

and (9.19.1) follows. □

9.3 Calculation of the Bergman Kernel for the Disc

Capsule: In general it is quite difficult to calculate the Bergman kernel (although there are ways to obtain asymptotic expansions for the kernel). The kernel can be related to the Green's function, and that gives a means of getting one's hands on the kernel.

For the disc, matters are simpler. Here we begin to explore the Bergman kernel for the unit disc D. We are able to give at least three different means of calculating the kernel in that context.

PROPOSITION 9.20 *The Bergman kernel for the unit disc D is*

$$K(z,\zeta) = \frac{1}{\pi} \cdot \frac{1}{(1 - z\bar{\zeta})^2}.$$

The Bergman metric for the disc is

$$g(z) = \frac{2}{(1-|z|^2)^2}.$$

This is (up to a constant multiple) the well-known Poincaré, or Poincaré-Bergman, metric.

This fact is so important that we now present three proofs. Some interesting function theory will occur along the way.

9.3.1 Construction of the Bergman Kernel for the Disc by Conformal Invariance

Let $D \subseteq \mathbb{C}$ be the unit disc. First we notice that, if either $f \in A^2(D)$ or $\overline{f} \in A^2(D)$, then

$$f(0) = \frac{1}{\pi} \iint_D f(\zeta)\, dA(\zeta). \tag{9.21}$$

This is the standard, two-dimensional area form of the mean value property for holomorphic or harmonic functions.

Of course the constant function $u(z) \equiv 1$ is in $A^2(D)$, so it is reproduced by integration against the Bergman kernel. Hence, for any $w \in D$,

$$1 = u(w) = \iint_D K(w, \zeta) u(\zeta)\, dA(\zeta) = \iint_D K(w, \zeta)\, dA(\zeta),$$

or

$$\frac{1}{\pi} = \frac{1}{\pi} \iint_D K(w, \zeta)\, dA(\zeta).$$

By (9.21), we may conclude that

$$\frac{1}{\pi} = K(w, 0)$$

for any $w \in D$.

Now, for $a \in D$ fixed, consider the Möbius transformation

$$h(z) = \frac{z - a}{1 - \overline{a}z}.$$

We know that

$$h'(z) = \frac{1 - |a|^2}{(1 - \overline{a}z)^2}.$$

We may thus apply Proposition 9.16 with $\phi = h$ to find that

$$\begin{aligned} K(w,a) &= h'(w) \cdot K(h(w),h(a)) \cdot \overline{h'(a)} \\ &= \frac{1-|a|^2}{(1-\overline{a}w)^2} \cdot K(h(w),0) \cdot \frac{1}{1-|a|^2} \\ &= \frac{1}{(1-\overline{a}w)^2} \cdot \frac{1}{\pi} \\ &= \frac{1}{\pi} \cdot \frac{1}{(1-w\overline{a})^2}. \end{aligned}$$

This is our formula for the Bergman kernel. The formula for the Bergman metric follows immediately by differentiation.

9.3.2 Construction of the Bergman Kernel by Means of an Orthonormal System

Now we will endeavor to write down the Bergman kernel for the disc by means of an orthonormal basis for $A^2(D)$, that is, by applying Proposition 9.13. For a general U, it can be rather difficult to actually write down a complete orthonormal system for $A^2(U)$. Fortunately, the unit disc $D \subseteq \mathbb{C}$ has enough symmetry that we can actually pull this off.

It is not difficult to see that $\{z^j\}_{j=0}^{\infty}$ is an *orthogonal* system for $A^2(D)$. That is, the elements are pairwise orthogonal; one sees this just by invoking polar coordinates. But they are not normalized to have unit length. We may confirm the first of these assertions by noting that, if $j \neq k$, then

$$\begin{aligned} \langle z^j, z^k \rangle &= \iint_D z^j \overline{z^k}\, dA(z) = \int_0^1 \int_0^{2\pi} r^j e^{ij\theta} r^k e^{-ik\theta}\, d\theta\, r dr \\ &= \int_0^1 r^{j+k+1}\, dr \int_0^{2\pi} e^{i(j-k)\theta}\, d\theta = 0. \end{aligned}$$

The system $\{z^j\}$ is complete: if $\langle f, z^j \rangle = 0$ for every j, then f will have a null power series expansion and hence be identically zero. It remains to normalize these monomials so that we have a complete ortho*normal* system.

We calculate that

$$\begin{aligned}\iint_D |z^j|^2\, dA(z) &= \int_0^1 \int_0^{2\pi} r^{2j}\, d\theta\, r dr \\ &= 2\pi \int_0^1 r^{2j+1}\, dr \\ &= \pi \cdot \frac{1}{j+1}.\end{aligned}$$

We conclude that

$$\|z^j\| = \frac{\sqrt{\pi}}{\sqrt{j+1}}.$$

Therefore the elements of our orthonormal system are

$$\phi_j(z) = \frac{\sqrt{j+1} \cdot z^j}{\sqrt{\pi}}.$$

Now, according to Proposition 9.13, the Bergman kernel is given by

$$\begin{aligned}K(z, \zeta) &= \sum_{j=0}^{\infty} \phi_j(z) \cdot \overline{\phi_j(\zeta)} \\ &= \sum_{j=0}^{\infty} \frac{(j+1) z^j \overline{\zeta^j}}{\pi} \\ &= \frac{1}{\pi} \cdot \sum_{j=0}^{\infty} (j+1) \cdot (z\overline{\zeta})^j.\end{aligned}$$

Observe that, in this instance, the convergence of the series is manifest (for both $|z| < 1$ and $|\zeta| < 1$).

Of course we easily can sum $\sum_j (j+1)\alpha^j$ by noticing that

$$\begin{aligned}\sum_{j=0}^{\infty} (j+1)\alpha^j &= \frac{d}{d\alpha} \sum_{j=0}^{\infty} \alpha^{j+1} \\ &= \frac{d}{d\alpha} \left[\alpha \cdot \frac{1}{1-\alpha} \right] \\ &= \frac{1}{(1-\alpha)^2}.\end{aligned}$$

Applying this result to our expression for $K(z, \zeta)$ yields that

$$K(z, \zeta) = \frac{1}{\pi} \cdot \frac{1}{(1 - z\overline{\zeta})^2}.$$

This is consistent with the formula that we obtained by conformal invariance in Subsection 9.3.1 for the Bergman kernel of the disc. The formula for the Bergman metric follows immediately by differentiation.

9.3.3 Construction of the Bergman Kernel by way of Differential Equations

It is actually possible to obtain the Bergman kernel of a domain in the plane from the Green's function for that domain (see [EPS]). Let us now summarize the key ideas. The reference [KRA2, Chapter 1] contains more detailed information about the Green's function. Unlike the first two Bergman kernel constructions, the present one will work for *any* domain with C^2 boundary.

First, the fundamental solution for the Laplacian in the plane is the function

$$\Gamma(\zeta, z) = \frac{1}{2\pi} \log |\zeta - z|.$$

This means that $\triangle_\zeta \Gamma(\zeta, z) = \delta_z$. [Observe that δ_z denotes the Dirac "delta mass" at z and $\triangle_\zeta$ is the Laplacian in the ζ variable.] Here the derivatives are interpreted in the sense of distributions. In more prosaic terms, the condition is that

$$\int \Gamma(\zeta, z) \cdot \triangle\varphi(\zeta)\, d\xi d\eta = \varphi(z)$$

for any C^2 function φ with compact support. We write, as usual, $\zeta = \xi + i\eta$.

Given a domain $U \subseteq \mathbb{C}$, the *Green's function* is posited to be a function $G(\zeta, z)$ that satisfies

$$G(\zeta, z) = \Gamma(\zeta, z) - F_z(\zeta),$$

where $F_z(\zeta) = F(\zeta, z)$ is a particular harmonic function in the ζ variable. Moreover, it is mandated that $G(\,\cdot, z)$ vanish on the boundary of U. One constructs the function $F(\,\cdot, z)$, for each fixed z, by solving a suitable Dirichlet problem. Again, the reference [KRA2, page 40] has all the particulars. It is worth noting, and this point will be discussed later, that the Green's function is a symmetric function of its arguments.

The next proposition establishes a striking connection between the Bergman kernel and the classical Green's function.

PROPOSITION 9.22 *Let $U \subseteq \mathbb{C}$ be a bounded domain with C^2 boundary. Let $G(\zeta, z)$ be the Green's function for U, and let $K(z, \zeta)$ be the Bergman kernel for U. Then*

$$K(z,\zeta) = 4 \cdot \overline{\frac{\partial^2}{\partial\zeta\partial\overline{z}} G(\zeta, z)}. \tag{9.22.1}$$

Proof: Our proof will use a version of Stokes's theorem written in the notation of complex variables. It says that, if $u \in C^1(\overline{U})$, then

$$\oint_{\partial U} u(\zeta)\, d\zeta = 2i \cdot \iint_U \frac{\partial u}{\partial \overline{\zeta}}\, d\xi\, d\eta, \tag{9.22.2}$$

where again $\zeta = \xi + i\eta$. The reader is invited to convert this formula to an expression in ξ and η and to confirm that the result coincides with the standard real-variable version of Stokes's theorem that can be found in any calculus book (see, e.g., [THO] and [BLK]).

Now we already know that

$$G(\zeta, z) = \frac{1}{4\pi} \log(\zeta - z) + \frac{1}{4\pi} \log \overline{(\zeta - z)} + F(\zeta, z). \tag{9.22.3}$$

Here we think of the logarithm as a multivalued holomorphic function; after we take a derivative, the ambiguity (which comes from an additive multiple of $2\pi i$) goes away.

Differentiating with respect to z (and using subscripts to denote derivatives), we find that

$$G_z(\zeta, z) = \frac{1}{4\pi} \frac{-1}{\zeta - z} + F_z(\zeta, z).$$

We may rearrange this formula to read

$$\frac{1}{\zeta - z} = -4\pi \cdot G_z(\zeta, z) + 4\pi F_z(\zeta, z).$$

We know that G, as a function of ζ, vanishes on ∂U. Hence so does G_z. Let $f \in C^2(\overline{U})$ be holomorphic on U. It follows that the Cauchy formula

$$f(z) = \frac{1}{2\pi i} \oint_{\partial U} \frac{f(\zeta)}{\zeta - z}\, d\zeta$$

can be rewritten as

$$f(z) = -2i \oint_{\partial U} f(\zeta) F_z(\zeta, z)\, d\zeta.$$

Now we apply Stokes's theorem (in the complex form) to rewrite this last as

$$f(z) = 4 \cdot \iint_U (f(\zeta)F_z)_{\overline{\zeta}}(\zeta, z)\, d\xi\, d\eta,$$

where $\zeta = \xi + i\eta$. Since f is holomorphic and F is real-valued, we may conveniently write this last formula as

$$f(z) = 4 \cdot \iint_U f(\zeta)\overline{F_{\zeta\overline{z}}}(\zeta, z)\, d\xi\, d\eta.$$

Now formula (9.22.3) tells us that $F_{\zeta\overline{z}} = G_{\zeta\overline{z}}$. Therefore we have

$$f(z) = \iint_U f(\zeta) 4\overline{G_{\zeta\overline{z}}}(\zeta, z)\, d\xi\, d\eta. \tag{9.22.4}$$

With a suitable limiting argument, we may extend this formula from functions f that are holomorphic and in $C^2(\overline{U})$ to functions in $A^2(U)$.

It is straightforward now to verify that $4\overline{G_{\zeta\overline{z}}}$ satisfies the first three characterizing properties of the Bergman kernel, just by examining our construction. The crucial reproducing property is formula (9.22.4). Then it follows that

$$K(z, \zeta) = 4 \cdot \overline{\frac{\partial^2}{\partial\zeta\partial\overline{z}} G(\zeta, z)}.$$

That is the desired result. □

It is worth noting that the proposition we have just established gives a practical method for confirming the existence of the Bergman kernel—by relating it to the Green's function, whose existence is elementary.

Now let us use the proposition to calculate a Bergman kernel. Of course the Green's function of the unit disc D is

$$G(\zeta, z) = \frac{1}{2\pi}\log|\zeta - z| - \frac{1}{2\pi}\log|1 - \zeta\overline{z}|,$$

as a glance at any classical complex analysis text will tell us (see, for example, [AHL2] or [HIL]). Verify the defining properties of the Green's function for yourself.

With formula (9.22.1) in mind, we can make life a bit easier by writing

$$\begin{aligned} G(\zeta, z) = \frac{1}{4\pi} \log(\zeta - z) + \frac{1}{4\pi} \log(\overline{\zeta - z}) \\ - \frac{1}{4\pi} \log(1 - \zeta\bar{z}) - \frac{1}{4\pi} \log\left(\overline{1 - \zeta\bar{z}}\right). \end{aligned}$$

Here we think of the expression on the right as the concatenation of four multivalued functions, in view of the ambiguity of the logarithm function. This ambiguity is irrelevant for us because the derivative of the Green's function is still well defined (i.e., the derivative annihilates additive constants).

Now we readily calculate that

$$\frac{\partial G}{\partial \bar{z}} = \frac{1}{4\pi} \cdot \frac{-1}{\bar{\zeta} - \bar{z}} + \frac{1}{4\pi} \cdot \frac{\zeta}{1 - \zeta\bar{z}}$$

and

$$\frac{\partial^2 G}{\partial \zeta \partial \bar{z}} = \frac{1}{4\pi} \cdot \frac{1}{(1 - \zeta\bar{z})^2}.$$

In conclusion, we may apply Proposition 9.22 to see that

$$K(z, \zeta) = \frac{1}{\pi} \cdot \frac{1}{(1 - z \cdot \bar{\zeta})^2}.$$

This result is consistent with that obtained in the first two calculations (Subsections 9.3.1, 9.3.2). The Bergman metric, as before, is obtained by differentiation.

9.4 A New Application

Capsule: Here we give a new proof of the profound, and relatively recently discovered, fact that a conformal selfmap of a planar domain that has three fixed points must be the identity. Our proof of course uses the Bergman theory.

We now present a rather profound, and relatively new, application of invariant metrics on planar domains. This result was first discovered around 1978 by several different authors. The complete history may be found in [FKKM1], [FKKM2]. We give here an argument from that paper. The result may be attributed to Fisher/Franks [FIF], Leschinger [LES], and Maskit [MAS]. See also [KRA8].

Before we begin the proof, some remarks are in order. First we note that, if φ is an automorphism of the disc D that has just two fixed points, then φ is the identity. For we may as well assume that one of the fixed points is the origin. By Schwarz's lemma, we know immediately that φ is a rotation. Now one additional fixed point forces φ to be the identity. The domain $\Omega = \mathbb{C}$ has a similar property. All automorphisms have the form $z \mapsto az + b$, and two fixed points provide the algebraic data to determine a and b uniquely.

Things are a bit more interesting for an annulus, say

$$\mathcal{A} = \left\{ z \in \mathbb{C} : \frac{1}{2} < |z| < 2 \right\}.$$

The inversion mapping $z \mapsto 1/z$ has 1 and -1 as fixed points. And of course it is not the identity. Any third fixed point would (see the theorem below) force an automorphism to be the identity. The Riemann sphere is similar. Any automorphism is a linear fractional transformation, and it is easy to see (on purely algebraic grounds) that three fixed points determine the transformation uniquely.

For a Riemann surface of arbitrary genus, things are more complicated. It is not difficult to see that, if k is a positive integer, then there is a Riemann surface with an automorphism φ that is not the identity such that φ has k fixed points. We now provide a couple of quick examples.

Example 9.23 Consider the complex one-dimensional torus T generated from the lattice $\{1, i\}$. Let $\pi : \mathbb{C} \to T$ be the standard covering map. Then $z \mapsto -z$ on the complex plane generates an automorphism, say τ, on T. Now τ has *four* fixed points, which are

$$\pi(0), \quad \pi(1/2 + i/2), \quad \pi(1/2), \quad \pi(i/2).$$

Yet τ does not fix $\pi(\frac{1}{4})$, and so it is not the identity map.

Example 9.24 We now consider a two-holed torus. This manifold can be generated by a regular octagon centered at the origin of the Poincaré disc together with its reflections. Again, $z \mapsto -z$ generates a nontrivial automorphism of this Riemann surface. The number of fixed points is now *six*, coming from the center (the origin), the vertices, and the corresponding pairs of midpoints of the sides of the octagon.

It is now clear that one can obtain arbitrarily large numbers of fixed points just from among the compact Riemann surfaces. See [FKKM1], [FKKM2] for the details.

Prelude: This result, first proved in 1978 and now having a good many proofs, is surprisingly precise and insightful. One of the nicest proofs shows that any domain has an equivariant embedding so that its automorphism group is a subset of the general linear group.

THEOREM 9.25 *Let $U \subseteq \mathbb{C}$ be any bounded, planar domain and $\phi : U \to U$ a conformal map. If ϕ has three fixed points (i.e., $\phi(p) = p$ for three distinct points p), then ϕ is the identity mapping.*

What is striking about this last result is that the number of fixed points needed does not depend on the topology of the domain. Results in [FKKM1], [FKKM2] show how, in higher dimensions, the number of fixed points depends on more subtle ideas depending on the geometry of cut loci. But, in the plane, the number three works all the time.

Maskit's proof of this last theorem is particularly pleasing. He shows that one can find a new domain Ω', conformally equivalent to the original Ω, so that the automorphisms of Ω' are all linear fractional transformations. Of course it is obvious that three fixed points force a linear fractional transformation to be the identity. Maskit's proof relies on ideas coming from uniformization. The other proofs are also subtle. We provide here a proof that just uses elementary geometry and topology, and which makes it clear why the number three is important.

Proof of the Theorem: We must begin with some terminology. We take it for granted that the reader has at least an intuitive understanding of the concept of "geodesic" (see [KON] for background). Going by the book, a geodesic is defined by a differential equation. For our purposes here, we may think of a geodesic as a locally length-minimizing curve.

Let $x \in M$, where M is a Riemannian manifold. A point $y \in M$ is called a *cut point* of x if there are two or more length-minimizing geodesics from x to y in M. See Figure 9.2. We further use the following basic terminology and facts from Riemannian geometry. Let $\text{dis}(x, y)$ denote the metric distance from x to y. A geodesic $\gamma : [a, b] \mapsto M$ is called a *length-minimizing geodesic* (or alternatively, a *minimal geodesic*, or a *minimal connector*) from x to y if $\gamma(a) = x$, $\gamma(b) = y$, and $\text{dis}\,(x, y) = [\text{arc length of } \gamma]$. By the Hopf-Rinow theorem, any two points in a complete Riemannian manifold can be connected by a minimizing geodesic. If there is a smooth family of minimizing geodesics from x to y, then these two points are said to

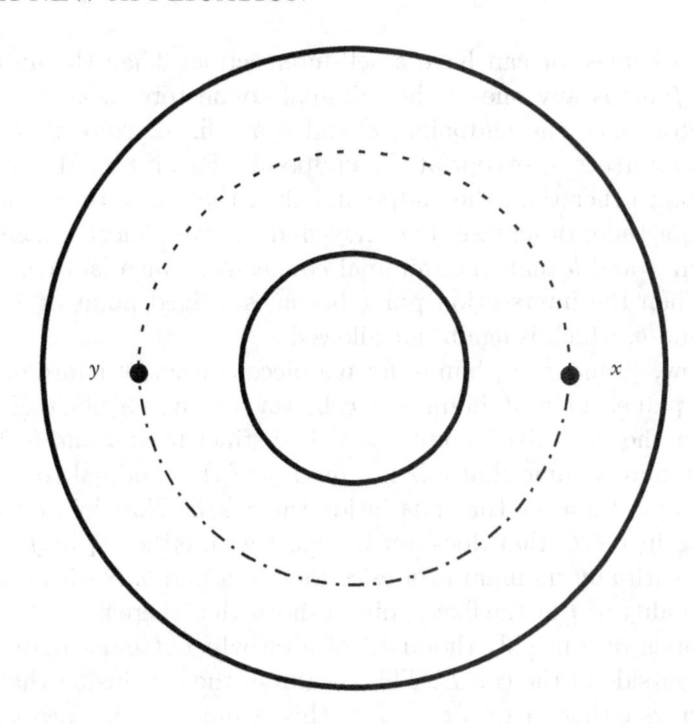

Figure 9.2: The point y is a cut point for x.

be *conjugate*. Conjugate points are cut points. The collection of cut points of x in M is called the *cut locus* of x, which we denote by C_x. It is known that C_x is nowhere dense in M (e.g., [GKM] and [KLI]); in fact, C_x lies in the singular set for the distance function.

Now equip U with the Bergman metric. For convenience, we shall suppose that U has C^1 boundary. This will guarantee that the Bergman metric is complete (see [OHS]). We assume that $f \in \text{Aut}\,(U)$ is not the identity map, but has three distinct fixed points in U. To reach a contradiction, let us start with the fixed point a. If the set of fixed points accumulates at a, we are done. So we may choose, as the second fixed point b, the closest (with respect to the Hermitian metric) one to a apart from a itself. This choice may not be unique, and hence we simply choose one.

We need to consider only the case that b is a cut point (not conjugate) of a. (For otherwise f would fix the unique geodesic connecting a to b and hence would be the identity.) Then there will be several unit-speed, minimal connectors (all of which have the same length, of course), say $\gamma_1, \gamma_2, \ldots$, joining a to b. First notice that no minimal connector can have a self-intersection. Then the automorphism f maps any one of the minimal connectors to another such connector since the endpoints a and b are fixed. Note that $f \circ \gamma_1$ cannot intersect γ_1 except at the endpoints. For, if they do intersect at a point other than the endpoints, then they have to intersect at the same time; otherwise, one may find an even shorter connector between a and b than the minimal connector, which is a contradiction. Then the intersection point becomes a fixed point of f closer to a than b, which is again not allowed.

Now, γ_1 and $f \circ \gamma_1$ join to form a piecewise smooth Jordan curve in the plane. Thus it bounds a cell, say E, in the plane $\mathbb{C}$. Now consider the third fixed point c that is distinct from a and b. Notice that we may assume that c is not on any of the minimal connectors for a and b. Suppose that c is inside the cell E. Now join c to a by an arc ξ in $E \cap U$ that does not intersect with either γ_1 or $f \circ \gamma_1$, or in fact with any minimal geodesics joining a and b. Notice that the conformality of f at the fixed point a shows that there is a sufficiently small open disc neighborhood W of a on which f must map $W \cap \xi$ to the outside of the cell E. This results in the conclusion that $f \circ \xi$ must cross either γ_1 or $f \circ \gamma_1$. But this is impossible, since a point not on any minimal connector from a to b cannot be mapped to a point on a minimal connector from a to b.

If c is outside the cell E, then the arguments are similar. Since there are only finitely many minimal connectors between a and b (since a and b are not conjugate to each other, and the quotient from the universal covering space is formed by a properly discontinuous group action; see [BRE]), some iterate f^m of f will move ξ so that its image has points inside E. Then, $f^m \circ \xi$ again crosses one of these minimal geodesics joining a and b, which leads us to another contradiction. □

Restricting attention back to the unit disc, we recall the difference between fixed points in the interior and fixed points in the boundary (this is a remark of Michael Christ). As we have noted, two fixed points in the interior force an automorphism to be the identity. But it takes three fixed points in the boundary for rigidity. One explanation for this difference is that two fixed points in the interior will give rise, by way of Schwarz reflection, to *four* fixed points of an automorphism of the Riemann sphere. That is impossible unless the map is the identity. But fixed points in the boundary will not multiply under reflection, so their behavior is different.

9.5 An Application to Mapping Theory

Capsule: Here we use an analysis of Bergman metric geodesics to classify all the conformal selfmaps of the annulus.

A classical result from complex function theory is the following.

Prelude: Here we study the automorphisms of an annulus using our new geometric insights. This proof is much more satisfying than the one usually found in textbooks.

THEOREM 9.26 *Let $\mathcal{A} = \{z \in \mathbb{C} : 1 < |z| < R\}$. If ϕ is a conformal mapping of $\mathcal{A}$ to itself, then ϕ is either a rotation or an inversion of the form $z \mapsto R/z$.*

It is interesting to note that it is quite difficult to compute the Bergman kernel and metric for an annulus. One can certainly see (using Laurent series) that the monomials $\{z^j\}_{j=-\infty}^{\infty}$ form an orthogonal system on an annulus centered at the origin. And one can use a little calculus to normalize these to an orthonormal system. But actually performing the necessary summation is virtually intractable (and involves elliptic functions [BER, pages 9–10]). Fortunately, the proof that we are about to present requires no detailed knowledge of the Bergman metric of the annulus. Indeed, it uses only the fact that the metric is complete, hence blows up at the boundary. We shall take this last result for granted, although see [APF] for a detailed treatment of this and related matters.

We will need to know that the Bergman metric has geodesics, but that follows from the smoothness and completeness of the metric (see [KON]). We will also utilize the real analyticity of the Bergman kernel and metric in an interesting and surprising way.

Proof of the Theorem: Again, we shall indulge the reader's intuition and invoke the heuristic idea of geodesic (see [KON] for all the details). Now fix an annulus $\mathcal{A} = \{z \in \mathbb{C} : 1 < |z| < R\}$. We will be using standard polar coordinates (r, θ). At any point $p \in \mathcal{A}$, a vector in the tangent space decomposes into a component in the $\partial/\partial r$ direction and a component in the $\partial/\partial\theta$ direction. One way to look at the metric is that it assigns a length to $\partial/\partial r$ and to $\partial/\partial\theta$ at each point of the annulus. Consider the set M of points where the Bergman length of $\partial/\partial\theta$ is minimal. Such points exist just because the Bergman metric blows up at the boundary of the domain. What geometric properties will the set M have?

First, the set must be rotationally invariant—because the metric will be rotationally invariant (i.e., the rotations are conformal self-maps of $\mathcal{A}$). Thus M is a union of circles centered at the origin. And M is certainly a closed set by the continuity of the metric. The set has no interior because the Bergman metric ρ is given by a real analytic function (it is the second derivative of the logarithm of the real analytic Bergman kernel), and the zero set of a nontrivial real analytic function can have no interior—see [KRP3]. In fact, we may also note that $|\partial/\partial\theta|_\rho$ is real analytic in the radial r direction; hence (by the same reasoning) we claim that M can have only finitely many circles in it. More precisely, let $g(r,\theta) = |\partial/\partial\theta|_{\rho,(r,\theta)}$. We have already noted that this function is independent of θ, so we may consider $g(r) = |\partial/\partial\theta|_{\rho,r}$. This function will assume its minimum value at points r where $g'(r) = 0$. Since the metric is complete, we know that the set of such points forms a compact subset of the interval $(0, R)$. If the set is infinite, then it has an accumulation point, and, therefore, the real analytic function g' is identically zero. That is impossible, again by the completeness of the metric. Thus the set M is finite.

Now it is easy to see that any curve (circle) in M will also be one of the curves (the curves that go once around the hole in the middle of the annulus) that minimizes arc length in the Bergman metric, and vice versa. This is so because we have already selected the curve to have $\partial/\partial\theta$ length as small as possible; a curve whose tangents have components in the $\partial/\partial r$ direction will a fortiori be longer. In more detail, the length of any curve $\gamma(t)$, $0 \leq t \leq 1$, is calculated by

$$\ell_\rho(\gamma) = \int_0^1 |\gamma'(t)|_{\rho,\gamma(t)}\,dt = \int_0^1 |\gamma'_r(t) + \gamma'_\theta(t)|_{\rho,\gamma(t)}\,dt,$$

where γ'_r and γ'_θ are, respectively, the normal and tangential components of γ'. Clearly, if we construct a new curve $\widetilde{\gamma}$ by integrating the vector field γ'_θ, then the result is a curve that is shorter than γ.

Thus any circle in M is a length-minimizing geodesic. And the converse is true as well.

Now let φ be a conformal self-map of $\mathcal{A}$. Then, as a result of the considerations in the last paragraph, φ will map circles in M (concentric with the annulus) to circles in M (concentric with the annulus). But more is true: *any* circle c that is centered at 0 has constant Bergman distance from any one given circle C in M. Let τ_c be the Bergman distance from the arbitrary circle c to the fixed circle C. Then c will be mapped by φ to another circle in $\mathcal{A}$ that has

Bergman distance τ_c from $\varphi(C)$. In short, φ maps circles to circles. And the same remark applies to φ^{-1}.

Now the orthogonal trajectories to the family of circles centered at P will be the radii of the annulus $\mathcal{A}$. Therefore (by conformality) these radii will get mapped to radii. Let $\mathcal{R}$ be the intersection of the positive real axis with the annulus $\mathcal{A}$. After composition with a rotation, we may assume that φ maps $\mathcal{R}$ to $\mathcal{R}$.

If M has just one circle in it, then that circle C must be fixed. Since each circle c in the annulus (centered at P) has a distance τ_c from C, then its image under φ will have the same distance from C. Thus either c is fixed or it is sent to its image under inversion. By continuity, whatever choice is valid for c will be valid for all other circles. Thus, in this case, the map φ is either the identity or inversion.

Now suppose that M contains at least two circles. The only possibilities for the action of φ on $M \cap \mathcal{R}$ are either preservation of the order of the points or inversion of the order (since any other permutation of the points is ruled out by elementary topology of a conformal mapping). By composing φ with an inversion ($z \mapsto R/z$), we may assume that φ does *not* act as an inversion on $\mathcal{R}$. But of course φ must preserve $M \cap \mathcal{R}$. We conclude that φ must be the identity.

What we have just proved is that, after we normalize φ so that it maps the positive real axis to itself, then φ must be the identity (or an inversion). In other words, the original map φ must be a rotation (or an inversion). $\square$

Chapter 10

A New Look at the Schwarz Lemma

Prologue: We have situated this new material on the Schwarz lemma at the end of the book for two reasons. One is that it is a dramatic way to end our discourse. A second is that it is motivated in part by considerations of Bergman geometry, a topic which we only covered in the preceding chapter.

We trust that the reader will find these closing ideas to be stimulating and rewarding, and will be motivated to pursue further reading in these topics.

10.1 The Boundary Schwarz Lemma

Capsule: A new idea that has arisen just in the past twenty-five years is to study the uniqueness part of the Schwarz lemma at a boundary point. This path of research was pioneered by D. Burns and S. G. Krantz among others. It turns out that such an investigation involves many interesting geometric and analytic ideas.

We treat this circle of thought in the present section.

There has been interest for some time in studying Schwarz lemmas at the boundary of a domain. Löwner conducted early studies of a result weaker than the one presented here; his motivation was the study of distortion theorems. Our methods are quite distinct from

Löwner's (see [VEL] for a consideration of classical results with references). The standard reference for this new material is the paper [BUK] of Burns and Krantz.

Prelude: This is the first fundamental result of Burns and Krantz. They were able to generalize it to a fairly large class of domains in several complex variables. People continue to study these ideas today.

THEOREM 10.1 *Let $\varphi : D \to D$ be a holomorphic function such that*

$$\varphi(\zeta) = 1 + (\zeta - 1) + \mathcal{O}(\|\zeta - 1\|^4)$$

as $\zeta \to 1$. Then $\varphi(\zeta) \equiv \zeta$ on the disc.

Compare this result with the uniqueness part of the classical Schwarz lemma. In that context, we assume that $\varphi(0) = 0$ and $\|\varphi'(0)\| = 1$. At the boundary, we must work harder. In fact, the following example shows that the size of the error term cannot be reduced from fourth order to third order:

Example 10.2 The function

$$\varphi(\zeta) = \zeta - \frac{1}{10}(\zeta - 1)^3$$

satisfies the hypotheses of the theorem with the exponent four replaced by three. Yet, clearly, this φ is not the identity.

To verify this example, we need only check that φ maps the disc to the disc. It is useful to let $\zeta = 1 - \tau$ and consider therefore the function

$$\widetilde{\varphi}(\tau) = 1 - \tau + \left[\frac{1}{10}\tau^3\right].$$

Now the result is clear by inspection. If $\arg \tau$ is near to 0, then the expression in brackets cannot push $\widetilde{\varphi}(\tau)$ past the edge of the disc. If $\arg \tau$ is larger, then one checks by hand that the expression in brackets has argument that causes it to push the value of the function *into* the disc. We leave details to the interested reader.

Proof of the Theorem: Consider the holomorphic function

$$g(\zeta) = \frac{1 + \varphi(\zeta)}{1 - \varphi(\zeta)}.$$

Then g maps the unit disc D to the right half-plane. By the Herglotz representation theorem,[1] there must be a positive measure μ on the interval $[0, 2\pi)$ and an imaginary constant $\mathcal{C}$ such that

$$g(\zeta) = \frac{1}{2\pi} \int_0^{2\pi} \frac{e^{i\theta} + \zeta}{e^{i\theta} - \zeta} \, d\mu(\theta) + \mathcal{C}. \tag{10.1.1}$$

We use the hypothesis on φ to analyze the structure of g and hence that of μ. To wit,

$$\begin{aligned} g(\zeta) &= \frac{1 + \zeta + \mathcal{O}(\zeta - 1)^4}{1 - \zeta - \mathcal{O}(\zeta - 1)^4} \\ &= \frac{1}{1 - \zeta} \cdot \frac{(1 + \zeta) + (\zeta - 1)^4}{1 - \mathcal{O}(\zeta - 1)^3} \\ &= \frac{1}{1 - \zeta} \cdot \left[(1 + \zeta) + \mathcal{O}(\zeta - 1)^4\right] \cdot \left[1 + \mathcal{O}(\zeta - 1)^3\right] \\ &= \frac{1 + \zeta}{1 - \zeta} + \mathcal{O}(\zeta - 1)^2. \end{aligned}$$

From this and equation (10.1.1) we easily conclude that the measure μ has the form $\delta_0 + \nu$, where δ_0 is (2π times) the Dirac mass at the origin and ν is another positive measure on $[0, 2\pi)$. (In fact, a nice way to verify the positivity of ν is to use the equation

$$\frac{1 + \zeta}{1 - \zeta} + \mathcal{O}(\zeta - 1)^2 = \frac{1}{2\pi} \int_0^{2\pi} \frac{e^{i\theta} + \zeta}{e^{e\theta} - \zeta} \, d(\delta_0 + \nu)(\theta) + \mathcal{C} \tag{10.1.2}$$

to derive a Fourier-Stieltjes expansion of $\delta_0 + \nu$ and then to apply the Herglotz criterion for the expansion of a positive measure (see [KAT, page 38]).)

We may simplify equation (10.1.2) to

$$\mathcal{O}(\zeta - 1)^2 = \frac{1}{2\pi} \int_0^{2\pi} \frac{e^{i\theta} + \zeta}{e^{i\theta} - \zeta} \, d\nu(\theta) + \mathcal{C}.$$

Now pass to the real part of the last equation, thereby eliminating the constant $\mathcal{C}$. Since ν is a positive measure, we see that the

[1]The reference [AHL3] has a thorough treatment of Herglotz's result. The reader may find it convenient to think of the result in this way. For $0 < r < 1$, the functions $G_r(e^{i\theta}) = \operatorname{Re} g(re^{i\theta})$ are all positive and all have mean value $g(0)$. Thus they form a bounded set in $L^1[0, 2\pi) \subseteq \mathcal{M}([0, 2\pi))$. Here $\mathcal{M}$ stands for the space of finite Borel measure, which is certainly the dual of the continuous functions. By the Banach-Alaoglu theorem (see [RUD3]), there is a subsequence G_{r_j} that converges weak-$*$ to a limit measure μ. This measure must be positive, and it is the measure that we seek. The Cauchy integral of μ gives a holomorphic function with the same real part as g. That real part determines the imaginary part up to an additive imaginary constant $\mathcal{C}$.

real part of the integral on the right represents a positive harmonic function h on the disc that satisfies

$$h(\zeta) = \mathcal{O}(\zeta - 1)^2. \tag{10.1.3}$$

In particular, h takes a minimum at the point $\zeta = 1$ and is $\mathcal{O}(\zeta - 1)^2$ there as well. This contradicts Hopf's lemma (Lemma 10.19 below) unless $h \equiv 0$. More specifically, Hopf's lemma tells us that a positive harmonic function with a minimum at the boundary point 1 must have nonvanishing normal derivative at 1. However, line (10.1.3) contradicts that assertion.

The only possible conclusion is that $\nu \equiv 0$, so that $\mu = \delta_0$. As a result,

$$g(\zeta) = \frac{1+\zeta}{1-\zeta}.$$

In conclusion, $\varphi(\zeta) \equiv \zeta$. The theorem is proved. □

Now we will treat some versions of the boundary Schwarz lemma that are due to Robert Osserman [OSS]. One of the remarkable features of Osserman's work is that he directly relates the interior Schwarz lemma to the boundary Schwarz lemma.

LEMMA 10.3 *Let $f : D \to D$ be a holomorphic function on the disc such that $f(0) = 0$. Then*

$$\|f(z)\| \le \|z\| \cdot \frac{\|z\| + \|f'(0)\|}{1 + \|f'(0)\|\|z\|}. \tag{10.3.1}$$

Proof: Set $g(z) = f(z)/z$. Then the standard Schwarz lemma tells us that either $\|g(z)\| < 1$ for all $z \in D$ or else f is a rotation. In the latter case, $\|f'(0)\| = 1$ and (10.3.1) holds trivially. Thus we need only consider the case $\|g(z)\| < 1$.

Since inequality (10.3.1) is unaffected by rotations, we may as well suppose that $g(0) = f'(0) = a$, where $0 \le a < 1$. In this case, (10.3.1) is equivalent to

$$\|g(z)\| \le \frac{\|z\| + a}{1 + a\|z\|}, \tag{10.3.2}$$

with $a = g(0)$. This last inequality is our Proposition 5.47. In particular, g must map each disc $D(0, r)$ into the image of that disc under the linear fractional map

$$\varphi_{-a}(\zeta) = \frac{\zeta + a}{1 + a\zeta}.$$

This latter image is the Euclidean disc whose diameter is the interval

$$\left[\frac{a-r}{1-ar}, \frac{a+r}{1+ar}\right]$$

in the real axis.

In conclusion,

$$\|z\| = r \Rightarrow \|g(z)\| \leq \frac{a+r}{1+ar} = \frac{|z|+a}{1+a\|z\|}.$$

This estimate establishes (10.3.2), and that, in turn, proves (10.3.1). □

Now we have a version of the boundary Schwarz lemma.

PROPOSITION 10.4 *Let $f : D \to D$ be holomorphic, and suppose that $f(0) = 0$. Further assume that, for some $b \in \partial D$, f extends continuously to b, $\|f(b)\| = 1$, and also that $f'(b)$ exists. Then*

$$\|f'(b)\| \geq \frac{2}{1 + \|f'(0)\|}.$$

Proof: We use Lemma 10.3. If $b, c \in \partial D$ with $f(b) = c$, then

$$\left\| \frac{f(z) - c}{\|z\| - \|b\|} \right\| \geq \frac{1 - \|f(z)\|}{1 - \|z\|} \geq \frac{1 + \|z\|}{1 + \|f'(0)\|\|z\|}.$$

As $\|z\| \to 1$, the right-hand side tends to $2/[1 + |f'(0)|]$. This reasoning proves that

$$\liminf_{z_j \to b} \left\| \frac{f(z_j) - c}{\|z_j\| - \|b\|} \right\| \geq \liminf_{z_j \to b} \frac{1 - \|f(z_j)\|}{1 - \|z_j\|} \geq \frac{2}{1 + \|f'(0)\|}.$$

If we choose $z_j = (1 - 1/j)b$, then the result follows. □

Osserman's more general boundary Schwarz lemma is as follows:

Prelude: Osserman's version of the boundary Schwarz lemma has a different flavor from that of Burns/Krantz. It has intrinsic interest.

THEOREM 10.5 *Let $f : D \to D$ be a holomorphic function such that, for some $b \in \partial D$, f extends continuously to b, $\|f(b)\| = 1$,* *and $f'(b)$ exists. (Note that we do not assume that $f(0) = 0$.) Define*

$$F(z) = \frac{f(z) - f(0)}{1 - \overline{f(0)} f(z)}.$$

Then

$$\|f'(b)\| \geq \frac{2}{1 + \|F'(0)\|} \cdot \frac{1 - \|f(0)\|}{1 + \|f(0)\|}.$$

Proof: Certainly F satisfies the hypotheses of Lemma 10.3, so that

$$\|F'(b)\| \geq \frac{2}{1 + \|F'(0)\|}. \tag{10.5.1}$$

But it is easy to calculate that

$$F'(z) = f'(z) \cdot \frac{1 - \|f(0)\|^2}{[1 - \overline{f(0)} f(z)]^2}.$$

Observe that $\|f(b)\| = 1$ implies that

$$\|1 - \overline{f(0)} f(b)\| \geq 1 - \|\overline{f(0)} f(b)\| = 1 - \|f(0)\|,$$

hence

$$\|F'(b)\| = \|f'(b)\| \cdot \frac{1 - \|f(0)\|^2}{\|1 - \overline{f(0)} f(b)\|^2} \leq \|f'(b)\| \cdot \frac{1 + \|f(0)\|}{1 - \|f(0)\|}. \tag{10.5.2}$$

Now (10.5.1) and (10.5.2) yield the desired conclusion. □

An interesting corollary of Proposition 10.4 is the following.

COROLLARY 10.6 *Under the hypotheses of Proposition 10.4, we have*

$$\|f'(b)\| \geq 1$$

and

$$\|f'(b)\| > 1 \quad \text{unless } f \text{ is a rotation.}$$

Proof: These two inequalities follow from Proposition 10.4 and the classical Schwarz lemma. □

Another interesting corollary, sort of an integrated form of the boundary Schwarz lemma, is the next result.

COROLLARY 10.7 *Let f satisfy all the conditions of Proposition 10.4, except for the hypothesis about the boundary point b. Assume that f extends continuously to a boundary arc $C \subseteq \partial D$ with $|f(z)| = 1$ for $z \in C$. Then the length s of C and the length σ of $f(C)$ satisfy*

$$\sigma \geq \frac{2}{1 + \|f'(0)\|} \cdot s.$$

Proof: By the Schwarz reflection principle, f extends to be holomorphic on C and, therefore, automatically satisfies the condition of Proposition 10.4 at the boundary point b for every point of C. Thus the corollary is immediate from the proposition. □

10.2 Liouville's and Picard's Theorems

Capsule: The geometric point of view enables us to see that Liouville's and Picard's theorems can be put into the same context as the Schwarz lemma. We explore that idea in this section.

It turns out that curvature gives criteria for when there do or do not exist nonconstant holomorphic functions from a domain U_1 to a domain U_2. The most basic result along these lines is as follows

Prelude: It is nice to see Liouville's theorem put into a geometric context. One cannot help but think of Ahlfors's version of the Schwarz lemma.

THEOREM 10.8 *Let $U \subseteq \mathbb{C}$ be an open set equipped with a metric $\sigma(z)$ having the property that its curvature $\kappa_\sigma(z)$ satisfies*

$$\kappa_\sigma(z) \leq -B < 0$$

for some positive constant B and for all $z \in U$. Then any holomorphic function

$$f : \mathbb{C} \to U$$

must be constant.

Proof: For $\alpha > 0$ we consider the mapping

$$f : D(0, \alpha) \to U.$$

Here $D(0, \alpha)$ is the Euclidean disc with center 0 and radius α, equipped with the metric $\rho_\alpha^A(z)$ as in Section 5.3. Fix $A > 0$. Theorem 5.35 yields, for any fixed z and $\alpha > |z|$, that

$$f^*\sigma(z) \leq \frac{\sqrt{A}}{\sqrt{B}}\rho_\alpha^A(z).$$

Letting $\alpha \to +\infty$ yields

$$f^*\sigma(z) \leq 0$$

hence

$$f^*\sigma(z) = 0.$$

This can be true only if $f'(z) = 0$. Since z was arbitrary, we conclude that f is constant. □

An immediate consequence of Theorem 10.8 is Liouville's theorem:

Prelude: Now we can explicitly interpret Theorem 10.8 as entailing Liouville's theorem. The proof is interesting.

THEOREM 10.9 *Any bounded, entire function is constant.*

Proof: Let f be such a function. After multiplying f by a constant, we may assume that the range of f lies in the unit disc. However, the Poincaré metric on the unit disc has constant curvature -4. Thus Theorem 10.8 applies and f must be constant. □

REMARK 10.10 This argument does not apply if the range is, for example, $\mathbb{C}$—because the plane $\mathbb{C}$ does not support a metric of strictly negative curvature. We shall the refine the proof of Theorem 10.8 in the ensuing discussions.

Picard's theorem is a dramatic strengthening of Liouville's theorem. It says that the hypothesis "bounded" may be weakened considerably; yet the same conclusion may be drawn.

Let us begin our discussion with a trivial example.

Example 10.11 Let f be an entire function taking values in the set

$$S = \mathbb{C} \setminus \{x + i0 : 0 \leq x \leq 1\}.$$

Following f by the mapping

$$\varphi(z) = \frac{z}{z-1},$$

we obtain an entire function $g = \varphi \circ f$ taking values in $\mathbb{C}$ less the set $\{x + i0 : x \leq 0\}$. If $r(z)$ is the principal branch of the square root function, then $h(z) = r \circ g(z)$ is an entire function taking values in the right half-plane. Now the Cayley map

$$c(z) = \frac{z-1}{z+1}$$

takes the right half-plane to the unit disc. So $u(z) = c \circ h(z)$ is a bounded, entire function. We conclude from Theorem 10.9 that u is constant. Unraveling our construction, we have that f is constant.

The point of this easy example is that, far from being bounded, an entire function need only omit a segment from its values in order that it be forced to be constant. And a small modification of the proof shows that the segment can be arbitrarily short. Picard considered the question of how small a set the image of a nonconstant, entire function can omit.

Let us pursue the same line of inquiry rather modestly by asking whether a nonconstant, entire function can omit one complex value. The answer is "yes," for $f(z) = e^z$ assumes all complex values except zero. It also turns out that it is impossible to construct a metric on the plane less a point that has negative curvature bounded away from zero.

The next step is to ask whether a nonconstant, entire function f can omit two values. The striking answer, discovered by Picard, is "no." Because of Theorem 10.8, it suffices for us to prove the following.

Prelude: This next theorem captures the geometric essence of the Picard theorem. The classical proof uses modular functions in a clever fashion. This proof uses curvature instead.

THEOREM 10.12 *Let U be a planar open set such that $\mathbb{C} \setminus U$ contains at least two points. Then U admits a metric μ whose curvature $\kappa_\mu(z)$ satisfies*

$$\kappa_\mu(z) \leq -B < 0$$

for some positive constant B.

Proof: After applying an invertible, affine map to U we may take the two omitted points to be 0 and 1 (if there are more than two omitted points, we may ignore the extras). Thus we will construct a metric of strictly negative curvature on $\mathbb{C}_{0,1} \equiv \mathbb{C} \setminus \{0, 1\}$.

Define

$$\mu(z) = \left[\frac{(1+\|z\|^{1/3})^{1/2}}{\|z\|^{5/6}}\right] \cdot \left[\frac{(1+\|z-1\|^{1/3})^{1/2}}{\|z-1\|^{5/6}}\right].$$

(After the proof, we will discuss where this nonintuitive definition came from.) The function μ is positive and smooth on $\mathbb{C}_{0,1}$. We proceed to calculate the curvature of μ.

First notice that, away from the origin,

$$\Delta\left(\log\|z\|^{5/6}\right) = \frac{5}{12}\Delta\left(\log\|z\|^2\right) = 0.$$

Thus

$$\begin{aligned}\Delta \log\left[\frac{(1+\|z\|^{1/3})^{1/2}}{\|z\|^{5/6}}\right] &= \frac{1}{2}\Delta \log\left(1+\|z\|^{1/3}\right) \\ &= 2\frac{\partial}{\partial z \partial \bar{z}} \log\left(1+[z\cdot\bar{z}]^{1/6}\right).\end{aligned}$$

Now a straightforward calculation leads to the identity

$$\Delta \log\left[\frac{(1+\|z\|^{1/3})^{1/2}}{\|z\|^{5/6}}\right] = \frac{1}{18}\frac{1}{\|z\|^{5/3}(1+\|z\|^{1/3})^2}.$$

The very same calculation shows that

$$\Delta \log\left[\frac{(1+\|z-1\|^{1/3})^{1/2}}{\|z-1\|^{5/6}}\right] = \frac{1}{18}\frac{1}{\|z-1\|^{5/3}(1+\|z-1\|^{1/3})^2}.$$

The definition of curvature now yields that

$$\begin{aligned}\kappa_\mu(z) = &-\frac{1}{18}\left[\frac{\|z-1\|^{5/3}}{(1+\|z\|^{1/3})^3(1+\|z-1\|^{1/3})}\right. \\ &\left. + \frac{\|z\|^{5/3}}{(1+\|z\|^{1/3})(1+\|z-1\|^{1/3})^3}\right].\end{aligned}$$

We record the following obvious facts:

(a) $\kappa_\mu(z) < 0$ for all $z \in \mathbb{C}_{0,1}$.

(b) $\lim_{z\mapsto 0} \kappa_\mu(z) = -\frac{1}{36}$.

(c) $\lim_{z \mapsto 1} \kappa_\mu(z) = -\frac{1}{36}$.

(d) $\lim_{z \mapsto \infty} \kappa_\mu(z) = -\infty$.

It follows immediately that κ_μ is bounded from above by a negative constant. □

REMARK 10.13 Now we discuss the motivation for the construction of μ. On looking at the definition of μ, one sees that the first factor is singular at 0 and the second is singular at 1. Let us concentrate on the first of these.

Since the expression defining curvature is rotationally invariant, it is plausible that the metric we define would also be rotationally invariant. Thus it should be a function of $\|z\|$. So one would like to choose an α such that $\|z\|^\alpha$ defines a metric of negative curvature. However, a calculation reveals that the α that is suitable for z large is not suitable for z small and vice versa. This explains why the expression has powers both of $\|z\|$ (for behavior near 0) and of $(1 + \|z\|)$ (for behavior near ∞). A similar discussion applies to the factors $\|z - 1\|^\alpha$.

Our calculations thus lead us to design the metric so that it behaves like $\|z\|^{-4/3}$ near ∞ and behaves like $\|z\|^{-5/6}$ (respectively $\|z - 1\|^{-5/6}$) near 0 (respectively 1).

We formulate Picard's little theorem as a corollary of Theorem 10.8 and Theorem 10.12:

COROLLARY 10.14 *Let f be an entire analytic function taking values in a set U. If $\mathbb{C} \setminus U$ contains at least two points, then f is constant.*

Proof: Since $\mathbb{C} \setminus U$ contains at least two points, Theorem 10.12 says that there is a metric of strictly negative curvature on U. But then Theorem 10.8 implies that any entire function taking values in U is constant. □

Entire functions are of two types: there are polynomials and non-polynomials *(transcendental* entire functions). Notice that a polynomial has a pole at infinity. Conversely, any entire function with a pole at infinity is a polynomial. So a transcendental entire function cannot have a pole at infinity and, being unbounded (by Liouville), cannot have a removable singularity at infinity. It therefore has an essential singularity at infinity.

Now notice that, by the fundamental theorem of algebra, a polynomial assumes all complex values. For a transcendental function, we analyze its essential singularity at infinity by recalling the Casorati-Weierstrass theorem: if 0 is an essential singularity for a holomorphic function f on a punctured disc $D'(0, \epsilon) \equiv D(0, \epsilon) \setminus \{0\}$, then f assumes values on $D'(0, \epsilon)$ that are *dense* in the complex plane. One might therefore conjecture that the essential feature of Picard's theorem is not that the function being considered is entire, but rather that it has an essential singularity at infinity. This is indeed the case and is the content of the great Picard theorem, which we now state:

Prelude: Picard's little theorem is a nice way of summarizing our ideas so far in this section.

THEOREM 10.15 *Let f be a holomorphic function on a punctured disc $D'(p, r) \equiv D(p, r) \setminus \{p\}$. Assume that f has an essential singularity at p. Then the restriction of f to any punctured disc $D'(p, s)$, $0 < s < r$, omits at most one value.*

10.3 Harmonic Functions

Capsule: Here we treat some basic ideas from harmonic function theory that will be needed in later sections.

Let $U \subseteq \mathbb{C}$ be a domain, that is, a connected open set. A twice continuously differentiable function u on U is said to be *harmonic* if it satisfies the partial differential equation

$$\triangle u \equiv \left(\frac{\partial^2}{\partial x^2} + \frac{\partial^2}{\partial y^2} \right) u(x, y) = 0 \tag{10.16}$$

at all points of U. The partial differential operator $\triangle$ is called the *Laplacian*, and the differential equation (10.16) is called *Laplace's equation*. We now review some of the key elementary ideas connected with this fundamental equation.

It is a matter of great interest to solve the "first boundary-value problem" associated to the Laplacian. This problem is also known as *Dirichlet's problem*. We quickly review its essential features. Let f be a continuous function on ∂U. We seek a function u that is

(i) continuous on $\overline{U}$,

(ii) twice continuously differentiable on U,

(iii) harmonic on U,

(iv) satisfies $u\big|_{\partial U} = f$.

Thus u is to be the harmonic continuation of f to the interior of U.

The Dirichlet problem has both mathematical and physical interest. If U is a thin metal plate, formed of heat-conducting material, and if f represents an initial heat distribution on the boundary of the plate, then the solution u of the Dirichlet problem represents the steady-state heat distribution in the interior.

It is worthwhile to take a moment and consider the special case when U is the unit disc $D = D(0,1)$. We may give a heuristic treatment of the Dirichlet problem as follows:

(a) In case $f(e^{i\theta}) = e^{ij\theta}$, $j = 0, 1, 2, \ldots$, then by inspection we see that $u(re^{i\theta}) = r^j e^{ij\theta}$, or $u(z) = z^j$, is the solution of the Dirichlet problem;

(b) In case $f(e^{i\theta}) = e^{ij\theta}$, $j = -1, -2, \ldots$, then by inspection we see that $u(re^{i\theta}) = r^{|j|} e^{ij\theta}$, or $u(z) = \overline{z}^{|j|}$, is the solution of the Dirichlet problem;

(c) In case $f(e^{i\theta}) = \sum_{j=-K}^{K} a_j e^{ij\theta}$, then, by linearity, the corresponding solution of the Dirichlet problem is

$$u(re^{i\theta}) = \sum_{j=0}^{K} a_j r^j e^{ij\theta} + \sum_{j=-K}^{-1} a_j r^{|j|} e^{ij\theta}.$$

Now the elementary theory of Fourier series (see [KRA3] or [RUD1]) tells us that any continuous function on the boundary of D may be uniformly approximated by trigonometric polynomials as in **(c)**. Thus we should be able to obtain the corresponding solution of the Dirichlet problem.

Proceeding formally, suppose that $f \sim \sum_{j=-\infty}^{\infty} \widehat{f}(j) e^{ij\theta}$, where

$$\widehat{f}(j) = \frac{1}{2\pi} \int_0^{2\pi} f(t) e^{-ijt}\, dt$$

is the standard Fourier coefficient (see [KRA4]). Then the corresponding solution of the Dirichlet problem should be

$$\begin{aligned} u(re^{i\theta}) &= \sum_{j=-\infty}^{\infty} \widehat{f}(j) r^{|j|} e^{ij\theta} \\ &= \sum_{j=-\infty}^{\infty} \frac{1}{2\pi} \int_0^{2\pi} f(t) e^{-ijt}\, dt \cdot r^{|j|} e^{ij\theta} \\ &= \frac{1}{2\pi} \int_0^{2\pi} f(t) \cdot \left[\sum_{j=-\infty}^{\infty} r^{|j|} e^{ij(\theta - t)} \right]. \end{aligned}$$

The expression in brackets is merely a double geometric series, and it is easily summed. What we find is that

$$u(re^{i\theta}) = \frac{1}{2\pi} \int_0^{2\pi} f(t) \cdot \frac{1 - r^2}{1 - 2r\cos(\theta - t) + r^2}\, dt. \qquad (10.17)$$

The expression

$$P(r, \theta) = \frac{1}{2\pi} \frac{1 - r^2}{1 - 2r\cos(\theta - t) + r^2}$$

is called the *Poisson kernel*, and the formula (10.17) is known as the *Poisson integral formula.*

For a general domain, one can almost never know the Poisson kernel explicitly. However, one can (with sufficiently powerful tools) obtain useful qualitative information. Here is a sample fact that is used frequently in the theory of partial differential equations and harmonic analysis.

PROPOSITION 10.18 *Let U be a bounded domain in $\mathbb{C}$ with C^2 boundary. Then the Poisson kernel $P(z, \zeta)$ for U satisfies*

$$c \cdot \frac{d(z)}{|z - \zeta|^2} \le P(z, \zeta) \le C \cdot \frac{d(z)}{|z - \zeta|^2}$$

for suitable positive constants c, C and for $d(z)$ the distance of z to ∂U.

Proof: See [KRA2] and [KRA4]. □

It is important also to observe that the Dirichlet problem may be solved for more general boundary data than a continuous function. In fact if the boundary function is L^∞, or even L^1, then its Poisson integral certainly makes sense. This fact will be crucial in what follows. Even without the Poisson integral (and the Poisson integral certainly will not exist in general on a domain with nonrectifiable boundary) then we can still consider the Dirichlet problem with, say, piecewise continuous boundary data. Such a Dirichlet problem is solved by the Perron method (see [GRK1, Ch. 7]).

10.4 Another Look at the Boundary Schwarz Lemma

Capsule: Here we take a look at Hopf's lemma, which is a primitive version of the boundary Schwarz lemma that is important both in partial differential equations and in the study of biholomorphic mappings.

10.4.1 Hopf's Lemma

The next result is one of the antecedents to a classical Schwarz lemma at the boundary. We shall first state the lemma, then discuss its context and significance.

LEMMA 10.19 (HOPF) *Let $U \subset\subset \mathbb{R}^N$ have C^2 boundary. Let $u \in C(\overline{U})$ be real-valued with u harmonic and non-constant on U. Let $P \in \partial U$ and assume that u takes a local minimum at P. Then*

$$\frac{\partial u}{\partial \nu}(P) < 0.$$

Proof: Suppose without loss of generality that $u > 0$ on U near P and that $u(P) = 0$. Let B_R be a ball that is internally tangent to ∂U at P. We may assume that the center of this ball is at the origin and that P has coordinates $(R, 0, \ldots, 0)$. Then, by Harnack's inequality (see [KR1]), we have for $0 < r < R$ that

$$u(r, 0, \ldots, 0) \geq c \cdot \frac{R^2 - r^2}{R^2 + r^2}$$

hence

$$\frac{u(r, 0, \ldots, 0) - u(R, 0, \ldots, 0)}{r - R} \leq -c' < 0.$$

Therefore

$$\frac{\partial u}{\partial \nu}(P) \leq -c' < 0.$$

This is the desired result. □

A good reference for the Hopf lemma is [COH]. It was used in that source to provide a proof of the maximum principal for second-order, elliptic partial differential operators. Namely, if a solution u of such an operator $\mathcal{L}$ has an interior maximum at a point P, then let S be a sphere passing through P. Restrict attention to the closed ball B bounded by S. Then the function u has a maximum at P, so the outward normal derivative at P is positive. But that means that, at a point near P in the outward normal direction the function u will take an even larger value, contradicting the maximality of u at P.

In more recent times the Hopf lemma has proved particularly useful in the study of biholomorphic and proper holomorphic mappings of several complex variables (see, for instance, [KRA2]).

The Hopf lemma is true for subharmonic functions, and under rather weak hypotheses on the behavior of u at P. We leave the details for the interested reader. The message that the Hopf lemma gives us is best seen for a holomorphic mapping $F : B \to B$, where B is the unit ball in $\mathbb{C}^n$. Let $\mathbf{1} = (1, 0, 0, \ldots, 0) \in \partial B$, and assume that the limit of $F(z)$ is $\mathbf{1}$ as z approaches $\mathbf{1}$ admissibly (see [KRA2] for this concept). Let ν be the outward unit normal vector to the boundary at $\mathbf{1}$, and set $f(z) = F(z) \cdot \nu$. Finally let $u(z) = |f(z)|$. Then u is plurisubharmonic, and u takes a maximum value (in a reasonable sense) at $\mathbf{1}$. The Hopf lemma applies, and we see that the normal derivative of u at $\mathbf{1}$ is nonzero. This tells us that the boundary point $\mathbf{1}$ is analytically isolated for the function f. And that is a primitive version of the Schwarz lemma at the boundary point $\mathbf{1}$.

10.5 Ideas of Löwner and Velling

Capsule: It turns out that C. Löwner (more than seventy years ago) and J. Velling (in his Ph.D. thesis) studied elementary versions of the boundary Schwarz lemma. Their point of view was to consider deformations.

As early as 1923, C. Löwner was considering deformation theorems that can be considered to be early versions of the Schwarz lemma at the boundary. A version of his result is this:

PROPOSITION 10.20 *Let $f : D \to D$ holomorphic with $f(0) = 0$. Of course f has radial boundary limits almost everywhere. Let $S = \partial D$. Assume that f maps an arc $A \subseteq S$ of length s onto an arc $f(A) \subseteq S$ of length σ. Then $\sigma \geq s$ with equality if and only if either $s = \sigma = 0$ or f is just a rotation.*

We see that Löwner tells us that a boundary arc must be mapped to a boundary arc that is shorter. This is in the spirit of the original Schwarz lemma, for it tells us that, under a similar mapping, the stretching factor must be less than 1.

John Velling studied Löwner's ideas in 1985 and proved a refinement which we shall treat at the end of the next section.

10.6 The Schwarz Lemma on the Boundary Redux

Capsule: Here we take another look at the boundary Schwarz lemma in the context of H. Cartan's classical uniqueness theorem for mappings.

It is worthwhile to formulate the classical Schwarz lemma in the language of the Burns/Krantz theorem. One way to do this is as follows.

LEMMA 10.21 *Let $f : D \to D$ be holomorphic, and assume that $f(0) = 0$. If*

$$f(\zeta) = \zeta + \mathcal{O}(|\zeta|^2), \tag{10.21.1}$$

then $f(\zeta) \equiv \zeta$.

The proof is obvious, for the hypothesis (10.21.1) implies that $f'(0) = 1$.

We might also recall H. Cartan's classic result.

Prelude: This result is a several-variable example of how our thoughts can be recast in higher dimensions. Note the role of the complex Jacobian.

THEOREM 10.22 *Let $U \subseteq \mathbb{C}^n$ be a bounded domain. Fix a point $P \in U$. Suppose that $\phi : U \to U$ is a holomorphic mapping such that $\phi(P) = P$. If the complex Jacobian of ϕ at P is the identity matrix, then ϕ is the identity mapping.*

We may think of Cartan's theorem as a reformulation of (10.21.1) in the multivariable setting. We now, for the sake of interest and completeness, provide a proof of Cartan's result.

Proof of Theorem 10.22: We may assume that $P = 0$. Expanding ϕ in a power series about $P = 0$ (and remembering that ϕ is vector-valued hence so is the expansion) we have

$$\phi(z) = z + P_k(z) + O(|z|^{k+1}),$$

where P_k is the first homogeneous polynomial of order exceeding 1 in the Taylor expansion. Defining $\phi^j(z) = \phi \circ \cdots \circ \phi$ (j times) we have

$$\begin{aligned} \phi^2(z) &= z + 2P_k(z) + O(|z|^{k+1}) \\ \phi^3(z) &= z + 3P_k(z) + O(|z|^{k+1}) \\ &\vdots \\ \phi^j(z) &= z + jP_k(z) + O(|z|^{k+1}). \end{aligned}$$

Choose polydiscs $D^n(0,a) \subseteq \Omega \subseteq D^n(0,b)$. Then, for $0 \le j \in \mathbb{Z}$, we know that $D^n(0,a) \subseteq \operatorname{dom} \phi^j \subseteq D^n(0,b)$. Therefore the Cauchy estimates imply that for any multi-index α with $|\alpha| = k$ we have

$$j|D^\alpha \phi(0)| = |D^\alpha \phi^j(0)| \le n \frac{b \cdot \alpha!}{a^k}.$$

Letting $j \to \infty$ yields that $D^\alpha \phi(0) = 0$.

We conclude that $P_k = 0$; this contradicts the choice of P_k unless $\phi(z) \equiv z$. □

REMARK 10.23 Notice that this proposition is a generalization of the uniqueness part of the classical Schwarz lemma on the disc. In fact a great deal of work has been devoted to generalizations of this type of Schwarz lemma to more general settings. We refer the reader to [WU], [YAU], [KRA4], [KRA5], [BUK] for more on this matter.

10.7 Chelst's Point of View

Capsule: Inspired by the work of Burns and Krantz, Chelst in his Ph.D. thesis developed some variants of the ideas. We treat those results, and also generalize them, here.

The following lemma is relevant to our considerations in this section.

LEMMA 10.24 *Let U be a bounded domain in $\mathbb{C}$ and let u be a real-valued harmonic function on U. Suppose that there is a collared neighborhood W of ∂U so that $u \ge 0$ on $U \cap W$. Then $u \ge 0$ everywhere.*

REMARK 10.25 It is not enough for u to simply be nonnegative on ∂U. As a simple example, let $\mathcal{U}$ be the upper-half-plane and let $u(x,y) = x^2 - y^2$. Then clearly $u \geq 0$ on $\partial\mathcal{U}$—indeed $u > 0$ at every point of $\partial\mathcal{U}$ except the origin. Yet u is *not* nonnegative on the positive imaginary axis.

Proof of the Lemma: Applying the maximum principle to $-u$ on a slightly smaller domain (with boundary lying inside $U \cap W$), we see that $-u$ cannot be positive in $U \setminus W$. Hence $u \geq 0$ on all of U. □

We will also make good use of the classical Hopf lemma, as enunciated in Lemma 10.19.

Now the following proposition is inspired by Chelst's main result, but is strictly more general. As a result, the line of argument is necessarily different.

PROPOSITION 10.26 *Let $f : D \to D$ be a holomorphic function. Let B be an inner function which equals 1 precisely on a set $A_B \subseteq \partial D$ of measure 0. Assume that:*

(a) *For a given point $a \in A_B$, $f(\zeta) = B(\zeta) + \mathcal{O}(|\zeta - a|^4)$ as $\zeta \to a$;*

(b) *For all $b \in A_B \setminus \{a\}$, $f(\zeta) = B(\zeta) + \mathcal{O}(|\zeta - b|^2)$ as $\zeta \to b$.*

Then $f(\zeta) \equiv B(\zeta)$ on all of D.

REMARK 10.27 It needs to be clearly understood here that A_B is the full set on which B equals 1. The proof consists of coming to terms with the boundary behavior of f and B on that set.

Proof: Following Chelst, it is useful to consider the function

$$h(\zeta) = \mathrm{Re}\left[\frac{1 + f(\zeta)}{1 - f(\zeta)}\right] - \mathrm{Re}\left[\frac{1 + B(\zeta)}{1 - B(\zeta)}\right].$$

We shall perform some estimates to show that **(i)** h has non-negative boundary limits almost everywhere on ∂D and **(ii)** h lies in $h^2(D)$ (i.e., harmonic functions which are uniformly square integrable on circles centered at the origin—see [KRA2]). The natural conclusion then is that h is positive everywhere on the interior of D.

Now

$$\begin{aligned}
h(\zeta) &= \mathrm{Re}\left[\frac{1+f(\zeta)}{1-f(\zeta)}\right] - \mathrm{Re}\left[\frac{1+B(\zeta)}{1-B(\zeta)}\right] \\
&= \mathrm{Re}\left[\frac{[1+B(\zeta)+\mathcal{O}(|\zeta-1|^4)]\cdot[1-\overline{B(\zeta)}+\mathcal{O}(|\zeta-1|^4)]}{|1-B(\zeta)+\mathcal{O}(|\zeta-1|^4)|^2}\right] \\
&\qquad - \mathrm{Re}\left[\frac{(1+B(\zeta))(1-\overline{B(\zeta)})}{|1-B(\zeta)|^2}\right] \\
&= \mathrm{Re}\left[\frac{(1-\overline{B(\zeta)}+B(\zeta)-|B(\zeta)|^2+\mathcal{O}(|\zeta-1|^4)}{|1-B(\zeta)+\mathcal{O}(|\zeta-1|^4)|^2}\right] \\
&\qquad - \mathrm{Re}\left[\frac{(1-\overline{B(\zeta)}+B(\zeta)-|B(\zeta)|^2}{|1-B(\zeta)|^2}\right] \\
&= \frac{[1-|B(\zeta)|^2+\mathcal{O}(|\zeta-1|^4)]\cdot|1-B(\zeta)|^2}{|1-B(\zeta)+\mathcal{O}(|\zeta-1|^4)|^2\cdot|1-B(\zeta)|^2} \\
&\qquad - \frac{[|1-B(\zeta)+\mathcal{O}(|\zeta-1|^4)]^2\cdot(1-|B(\zeta)|^2)}{|1-B(\zeta)+\mathcal{O}(|\zeta-1|^4)|^2\cdot|1-B(\zeta)|^2} \\
&= \frac{[(1-|B(\zeta)|^2)\cdot|1-B(\zeta)|^2+\mathcal{O}(|\zeta-1|^4]}{|1-B(\zeta)|^4} \\
&\qquad - \frac{[|1-B(\zeta)|^2\cdot(1-|B(\zeta)|^2)+\mathcal{O}(|\zeta-1|^4)]}{|1-B(\zeta)|^4} \\
&= \frac{\mathcal{O}(|\zeta-1|^4)}{|1-B(\zeta)|^4}.
\end{aligned}$$

But Hopf's lemma tells us that $|1-B(\zeta)|$ is *not* $o(|\zeta-1|)$. And in fact we can certainly say (a bit sloppily) that $|1-B(\zeta)| \geq C\cdot|1-\zeta|^{1+\epsilon}$ for some small $\epsilon > 0$.

In conclusion, the function h certainly lies in $h^2(D)$. We also note that (and our calculations show this) the boundary limits of the first expression on the right-hand side of the first line of the previous multi-line display are nonnegative almost everywhere. And the boundary limits of the second expression on the right-hand side of the first line of the previous multi-line display are 0 almost everywhere. In summary, we have an h^2 harmonic function with nonnegative radial boundary limits almost everywhere. It then follows, from the Poisson integral formula for instance, that h is positive on the disc D.

But h takes the boundary limit 0 at each point of A_B. It follows then from Hopf's lemma that h has a nonzero normal derivative at each of those points. That fact contradicts hypothesis **(b)** of the proposition. And that contradiction tells us that $h \equiv 0$, hence f is identically equal to the Blaschke product B. $\square$

Chelst [CHE] has pointed out that the function

$$f(\zeta) = \zeta^8 - \frac{1}{256}(\zeta+1)\left[(\zeta^2+1)(\zeta^4+1)\right]^2 \cdot (\zeta-1)^4$$

maps the disc to the disc and fails hypothesis **(b)** of Proposition 10.26 with $A_B = \{-1, 1\}$ and $B = \zeta \cdot \zeta$; it also fails the conclusion.

It should be mentioned that the papers [BZZ] and [SHO] offer further refinements of the Burns/Krantz and Chelst theorems.

10.8 Several Complex Variables

Capsule: Although this is really a book about one complex variable, one can gain some perspective by looking at matters from the several-complex-variable point of view. Here we demonstrate this idea in the context of the Schwarz lemma.

The work from [BUK] described above in Section 10.1, in the one-complex-variable setting, was inspired by a question of several complex variables. Namely one wanted to know whether a holomorphic mapping $\Phi : B \to B$ (where B is the unit ball in $\mathbb{C}^n$) could have boundary image $\Phi(\partial B)$ with high order of contact with the target boundary ∂B. In one complex variable, the Riemann Mapping theorem tells us that, for a holomorphic mapping $\varphi : D \to D$, any order of contact of $\varphi(\partial D)$ with the target boundary ∂D is possible. Of course there is no Riemann Mapping theorem in several complex variables, and this together with other *ad hoc* evidence suggested that there ought to be an upper bound on the order of contact in the multi-dimensional case.

The first step in understanding this situation is to prove a multi-dimensional version of Theorem 10.1.

Prelude: This is the quintessential version of the Burns/Krantz theorem. It is a new boundary version of the uniqueness part of the Schwarz lemma in several complex variables.

THEOREM 10.28 *Let B be the unit ball in $\mathbb{C}^n$. Let $\Phi : B \to B$ be a holomorphic mapping. Let $\mathbf{1} \equiv (1, 0, 0, \ldots, 0)$ be the usual boundary point of the ball. Assume that*

$$\Phi(z) = \mathbf{1} + (z - \mathbf{1}) + \mathcal{O}(|z - \mathbf{1}|^4).$$

Then $\Phi(z) \equiv z$ for all $z \in B$.

Proof: For simplicity we restrict attention to complex dimension two. For each $a \in B$, let $\mathcal{L}_a$ be the complex line passing through a and $\mathbf{1}$. Let $\mathbf{d}_a$ be the complex disc given by $\mathcal{L}_a \cap B$. With a fixed, consider the holomorphic function

$$\begin{aligned} \psi : D &\longrightarrow B \\ \zeta &\longmapsto (\zeta, 0). \end{aligned}$$

Also consider the mapping

$$\phi_a : B \to B$$

which is the automorphism of the ball B which maps $\mathbf{d}_0$ onto $\mathbf{d}_a$ and fixes $\mathbf{1}$. Indeed one may say rather explicitly what this last automorphism is. Note that, for α a complex number of modulus less than 1, the mapping

$$\lambda_\alpha(z_1, z_2) = \begin{pmatrix} \dfrac{(1 - |\alpha|^2) z_1}{1 + \overline{\alpha} z_2} + \dfrac{\overline{\alpha}(z_2 + \alpha)}{1 + \overline{\alpha} z_2} \\[2ex] \dfrac{-\alpha\sqrt{1 - |\alpha|^2}\, z_1}{1 + \overline{\alpha} z_2} + \dfrac{(z_2 + \alpha)\sqrt{1 - |\alpha|^2}}{1 + \overline{\alpha} z_2} \end{pmatrix}$$

sends the complex line $\mathbf{d}_0$ through $(0, 0)$ and $(1, 0)$ to the complex line through $(|\alpha|^2, \alpha\sqrt{1 - |\alpha|^2})$. Composition with unitary mappings will allow us to replace $(|\alpha|^2, \alpha\sqrt{1 - |\alpha|^2})$ with any other element of the ball B.

Finally define

$$\begin{aligned} \pi_1 : B &\longrightarrow B \\ (z_1, z_2) &\longmapsto (z_1, 0) \end{aligned}$$

and

$$\begin{aligned} \eta : \mathbf{d}_0 &\longrightarrow D \\ (z_1, 0) &\longmapsto z_1. \end{aligned}$$

The function

$$\begin{aligned} H_a : D &\longrightarrow D \\ \zeta &\longmapsto \eta \circ \pi_1 \circ (\phi_a)^{-1} \circ \Phi \circ \phi_a \circ \psi(\zeta) \end{aligned}$$

is well defined. In addition, H satisfies the hypotheses of Theorem 10.1. It follows then that $H_a(\zeta) \equiv \zeta$.

Now set

$$G_a(\zeta) = (\phi_a)^{-1} \circ \Phi \circ \phi_a \circ \psi(\zeta) \equiv \big(g_a^1(\zeta), g_a^2(\zeta)\big).$$

The statement that $H_a(\zeta) \equiv \zeta$ tells us that $g_a^1(\zeta) \equiv \zeta$. But then

$$|g_a^1(\zeta)|^2 + |g_a^2(\zeta)|^2 < 1$$

for $\zeta \in D$.

Letting $|\zeta| \to 1$ now yields that $|g_a^2(\zeta)| \to 0$. Thus $g_a^2 \equiv 0$. It now follows that the image of G_a already lies in $\mathbf{d}_0$. Consequently it must be that Φ preserves $\mathbf{d}_a$. This last assertion can hold for every choice of a if and only if Φ is the identity mapping.

That completes the proof. □

It is naturally desirable to extend this last result to a more general class of domains. The key insight here is to note that the discs $\mathbf{d}_a$ in B may be replaced, in a more general setting, by extremal discs for the Kobayashi metric (see, for instance [KRA5] and especially [LEM]). The theory of such discs is well developed in the context of strongly convex domains, and the proof we have given here transfers naturally to that setting.

For strongly pseudoconvex domains, there is no theory of extremal discs in the sense of Lempert (but see [KRA5]). However, Burns and Krantz [BUK] were able to construct a local theory of extremal discs near a strongly pseudoconvex boundary point. As a result, it is possible to prove a version of Theorem 10.28 on a smoothly bounded, strongly pseudoconvex domain. Details may be found in [BUK]. For the record, we record the result now.

Prelude: Using Lempert's theory of extremal discs for the Kobayashi metric, Burns and Krantz were able to generalize their fundamental result to strongly pseudoconvex domains. Here geometry is playing a decisive role.

THEOREM 10.29 *Let $\Omega \subseteq \mathbb{C}^n$ be a smoothly bounded, strongly pseudoconvex domain. Let $\Phi : \Omega \to \Omega$ be a holomorphic mapping. Let $P \in \partial\Omega$ be a boundary point. Assume that*

$$\Phi(z) = P + (z - P) + \mathcal{O}(|z - P|^4).$$

Then $\Phi(z) \equiv z$ for all $z \in \Omega$.

We return now to the original question of order of contact of the image of a holomorphic mapping with the target of the mapping. Although we cannot provide all the technical details here (see [BUK] for the chapter and verse), we can at least state the definitive result of Burns and Krantz. First a definition.

Definition 10.30 Let B be the unit ball in $\mathbb{C}^n$ and $\Phi : B \to B$ a holomorphic mapping. We let ρ be a defining function for the domain ball and ρ' a defining function for the range ball. Let $P \in \partial B$ and $Q = \Phi(P) \in \partial B$ the image point. We say that $\Phi(\partial B)$ has *geometric contact* with ∂B of weight N at $\Phi(P) = Q$ if

$$\rho' \circ \Phi = h \cdot \rho + w_N,$$

where h is a smooth, positive function on $\overline{B}$ near P and w_N vanishes at P to order N in a suitable nonisotropic sense.

Prelude: This theorem answer the original question of Warren Wogen that motivated the work in [BUK].

THEOREM 10.31 *Let B be the unit ball in $\mathbb{C}^n$. Assume that $n \geq 2$. Let $\Phi : B \to B$ be a holomorphic mapping. Assume that Φ is C^6 to the boundary near a point $P \in \partial B$ and its image point $Q = \Phi(P) \in \partial B$. If $\Phi(\partial B)$ has geometric contact of weight six with ∂B at Q, then there is a global biholomorphism $\Psi : B \to B$ such that*

$$\Phi(z) = \Psi(z) + o(|z - P|^3)$$

near P.

We close this section by using Proposition 10.26 to derive a new version of Theorem 10.28.

Prelude: Here we have a variant of the boundary Schwarz lemma on the ball that is derived by methods that we have seen previously.

THEOREM 10.32 *Let $f : B \to B$ be a holomorphic function. Let h be an inner function which equals 1 on a set $A_h \subseteq \partial B$ of measure 0. Assume that*

(a) *For a given point* $a \in A_h$, $f(z) = B(z) + \mathcal{O}(|z-a|^4)$ *as* $z \to a$*;*

(b) *For all* $b \in A_h \setminus \{a\}$, $f(z) = B(z) + \mathcal{O}(|z-b|^2)$ *as* $z \to b$*.*

Then $f(z) \equiv h(z)$ *on all of* D*.*

Proof: As indicated. □

It is worth mentioning that the work in [FEF] shows that the hypothesis of Theorem 10.31 implies that the boundary asymptotics of the Bergman metric are preserved (asymptotically at P) by the mapping Φ. In particular, pseudo-transversal geodesics (in the language of Fefferman) are mapped to pseudo-transversal geodesics. And the asymptotic expansion for the Bergman kernel is mapped to itself in a natural way.

Bibliography

[AHL1] L. Ahlfors, *Conformal Invariants: Topics in Geometric Function Theory*, McGraw-Hill, New York, 1973.

[AHL2] L. Ahlfors, *Complex Analysis*, 3rd ed., McGraw-Hill, New York, 1976.

[AHL3] L. Ahlfors, *Lectures on Quasiconformal Mappings*, D. Van Nostrand Col, Princeton, NJ, 1966.

[AHL4] L. Ahlfors, Bounded analytic functions, *Duke Mathematical Journal* 14(1947), 1–11.

[ALK] G. Aladro and S. G. Krantz, A criterion for normality in $\mathbb{C}^n$, *Jour. Math. Anal. Appl.* 161(1991), 1–8.

[AND1] E. M. Andreev, On convex polyhedra in Lobačevskiĭ space, *Matematicheskii Sbornik* (N.S.) 81(1970), 445–478; English transl. in *Mathematics of the USSR–Sbornik* 12(1970), 413–440.

[AND2] E. M. Andreev, Convex polyhedra of finite volume in Lobačevskiĭ space, *Matematicheskii Sbornik* (N.S. 83(1970), 256–260; English transl. in *Mathematics of the USSR–Sbornik* 12(1970), 255–256.

[APF] L. Apfel, Localization properties and boundary behavior of the Bergman kernel, thesis, Washington University, 2003.

[BED] E. Bedford and J. Dadok, Bounded domains with prescribed group of automorphisms, *Commentarii Mathematici Helvetici* 62(1987), 561–572.

[BEL] J. Bergh and J. Löfström, *Interpolation Spaces: An Introduction*, Springer-Verlag, New York, 1973.

[BER] S. Bergman, *The Kernel Function and Conformal Mapping*, American Mathematical Society, Providence, RI, 1970.

[BIS] C. Bishop, *The Riemann Mapping Theorem*, manuscript, 368 pp.

[BLK] B. Blank and S. G. Krantz, *Calculus*, Key Curriculum Press, Emeryville, CA, 2006.

[BRO] F. Browder, *Mathematical Developments Arising from the Hilbert Problems*, American Mathematical Society, Providence, RI, 1976.

[BUR] R. Burckel, *An Introduction to Classical Complex Analysis*, Birkhäuser Publishing, Basel, 1976.

[BUK] D. Burns and S. G. Krantz, Rigidity of holomorphic mappings and a new Schwarz lemma at the boundary, *Journal of the American Mathematical Society* 7(1994), 661–676.

[CAL] A. P. Calderón, Intermediate spaces and interpolation, the complex method, *Studia Mathematica* 24(1964), 113–190.

[CIS1] J. A. Cima and T. J. Suffridge, Proper holomorphic mappings from the two-ball to the three-ball, *Transactions of the American Mathematical Society* 311(1989), 227–239.

[CIS2] J. A. Cima and T. J. Suffridge, Proper mappings between balls in $\mathbb{C}^n$, *Complex Analysis* (University Park, PA., 1986), 66–82, Lecture Notes in Mathematics, 1268, Springer, Berlin, 1987.

[DANG1] J. P. D'Angelo, The structure of proper rational holomorphic maps between balls, *Several complex variables* (Stockholm, 1987/1988), 227–244, Math. Notes, 38, Princeton Univ. Press, Princeton, NJ, 1993.

[DANG2] J. P. D'Angelo, The geometry of proper holomorphic maps between balls, *The Madison Symposium on Complex Analysis* (Madison, WI, 1991), 191–215, Contemporary Mathematics, 137, American Mathematical Society, Providence, RI, 1992.

[DANG3] J. P. D'Angelo, Real hypersurfaces, orders of contact, and applications, *Annals of Mathematics* 115(1982), 615–637.

[DER] G. de Rham, *Variétés Différentiable*, Springer Verlag, Berlin and New York, 1984.

[DIK] F. Di Biase and S. G. Krantz, *The Boundary Behavior of Holomorphic Functions*, Birkhäuser Publishing, Boston, MA, to appear.

[DIF] K. Diederich and J. E. Fornæss, Pseudoconvex domains: Bounded strictly plurisubharmonic exhaustion functions, *Inventiones Mathematicae* 39(1977), 129–141.

[FAK] H. Farkas and I. Kra, *Riemann Surfaces*, 2nd ed., Springer-Verlag, New York, 1992.

[FED] H. Federer, *Geometric Measure Theory*, Springer-Verlag, New York and Berlin, 1969.

[FEF] C. Fefferman, The Bergman kernel and biholomorphic mappings of pseudoconvex domains, *Inventiones Mathematicae* 26(1974), 1–65.

[FIS] S. Fisher, *Function Theory on Planar Domains*, John Wiley & Sons, New York, 1983.

[FIF] S. D. Fisher and John Franks, The fixed points of an analytic self-mapping, *Proceedings of the American Mathematical Society*, 99(1987), 76–78.

[FKKM1] B. Fridman, K.-T. Kim, S. G. Krantz, and D. Ma, On fixed points and determining sets for automorphisms, *Michigan Journal of Mathematics* 50(2002), 507–515.

[FKKM2] B. Fridman, K.-T. Kim, S. G. Krantz, and D. Ma, On determining sets for holomorphic automorphisms, *Rocky Mountain Journal of Mathematics* 36(2006), 947–955.

[GAM] T. Gamelin, *Complex Analysis*, Springer-Verlag, New York, 2001.

[GAR1] J. B. Garnett, *Analytic Capacity and Measure*, Springer Lecture Notes v. 297, Springer-Verlag, Heidelberg, 1972.

[GAR2] J. B. Garnett, *Bounded Holomorphic Functions*, Academic Press, New York, 1981.

[GARM] J. B. Garnett and D. Marshall, *Harmonic Measure*, Cambridge University Press, Cambridge, 2005.

[GOL] Goluzin, *Geometric Theory of Functions of a Complex Variable*, American Mathematical Society, Providence, RI, 1966.

[GRH] M. J. Greenberg and J. R. Harper, *Algebraic Topology*, Benjamin/Cummings, Reading, MA, 1981.

[GKK] R. E. Greene, K.-T. Kim, and S. G. Krantz, *The Geometry of Complex Domains*, Birkhäuser Publishing, Boston, MA, 2011.

[GRK1] R. E. Greene and S. G. Krantz, *Function Theory of One Complex Variable*, 3rd ed., American Mathematical Society, Providence, RI, 2006.

[GKM] D. Gromoll, W. Klingenberg, and W. Meyer, *Riemannsche Geometrie im Grossen*, 2nd ed., Lecture Notes in Mathematics, v. 55, Springer-Verlag, New York, 1975.

[HES] Z. He and O. Schramm, Koebe uniformization and circle packing, *Annals of Mathematics* 137(1993), 369–4013.

[HEI1] M. Heins, A note on a theorem of Radó concerning the $(1, m)$ conformal maps of a multiply-connected domain into itself, *Bulletin of the American Mathematical Society* 47(1941), 128–130.

[HEI2] M. Heins, On the number of 1–1 directly conformal maps which a multiply-connected plane domain of finite connectivity p (> 2) admits onto itself, *Bulletin of the American Mathematical Society* 52(1946), 454–457.

[HEL] S. Helgason, *Differential Geometry and Symmetric Spaces*, Academic Press, New York, 1962.

[HIL] E. Hille, *Analytic Function Theory*, 2nd ed., Ginn and Co., Boston, 1973.

[HOF] K. Hoffman, *Banach Spaces of Analytic Functions*, Prentice-Hall, Englewood Cliffs, NJ, 1962.

[HUA] X. Huang, A boundary rigidity problem for holomorphic mappings on some weakly pseudoconvex domains. *Canadian Journal of Mathematics* 47(1995), 405–420.

[HUS] D. Husemoller, *Fibre Bundles*, 2nd ed., Springer-Verlag, New York, 1975.

[KAT] Y. Katznelson, *An Introduction to Harmonic Analysis*, John Wiley & Sons, New York, 1968.

[KLI] W. Klingenberg, *Riemannian Geometry*, 2nd ed., de Gruyter Studies in Mathematics, Berlin, 1995.

[KOB] S. Kobayashi, *Hyperbolic Manifolds and Holomorphic Mappings*, Marcel Dekker, New York, 1970.

[KON] S. Kobayashi and K. Nomizu, *Foundations of Differential Geometry*, Vols. I and II, Interscience, New York, 1963, 1969.

[KOO] P. Koosis, *Introduction to H_p Spaces*, 2nd ed., Cambridge Tracts in Mathematics, v. 115, Cambridge University Press, Cambridge, 1998.

[KRA1] S. G. Krantz, *Complex Analysis: The Geometric Viewpoint*, 2nd ed., Mathematical Association of America, Washington, D.C., 2004.

[KRA2] S. G. Krantz, *Function Theory of Several Complex Variables*, 2nd ed., American Mathematical Society, Providence, Rhode Island, 2001.

[KRA3] S. G. Krantz, *A Panorama of Harmonic Analysis*, Mathematical Association of America, Washington, D.C., 1999.

[KRA4] S. G. Krantz, Calculation and estimation of the Poisson kernel, *Journal of Mathematical Analysis and Applications* 302(2005), 143–148.

[KRA5] S. G. Krantz, The Kobayashi metric, extremal discs, and biholomorphic mappings, *Complex Variables and Elliptic Equations*, 57(2012), 1–14.

[KRA6] S. G. Krantz, *Elements of Topology: Theory and Practice*, Taylor & Francis, Boca Raton, FL, 2009.

[KRA7] S. G. Krantz, *Cornerstones of Geometric Function Theory: Explorations in Complex Analysis*, Birkhäuser Publishing, Boston, MA, 2006.

[KRA8] S. G. Krantz, Two results on uniqueness of conformal mappings, *Complex Variables and Elliptic Equations* 50(2005), 427–432.

[KRP1] S. G. Krantz and H. R. Parks, *The Implicit Function Theorem*, Birkhäuser, Boston, 2002.

[KRP2] S. G. Krantz and H. R. Parks, *The Geometry of Domains in Space*, Birkhäuser, Boston, 1996.

[KRP3] S. G. Krantz and H. R. Parks, *A Primer of Real Analytic Functions*, 2nd ed., Birkhäuser, Boston, 2002.

[LEF] S. Lefschetz, *Algebraic Topology*, Princeton University Press, Princeton, NJ, 1953.

[LEM] L. Lempert, La metrique Kobayashi et las representation des domains sur la boule, *Bulletin de la Société Mathématique de France* 109(1981), 427–474.

[LES] K. Leschinger, Über fixpunkte holomorpher Automorphismen, *Manuscripta Math.*, 25(1978), 391–396.

[LOS] L. Loomis and S. Sternberg, *Advanced Calculus*, Addison-Wesley, Reading, MA, 1968.

[LUR] D. H. Luecking and L. A. Rubel, *Complex Analysis: A Functional Analysis Approach*, Springer-Verlag, New York, 1984.

[LYP] J. Lyons and J. Peetre, Sur une classe d'espaces d'interpolation, *Inst. Hautes Études Sci. Publ. Math.*, No. 19, 1964, 5–68.

[MAS] B. Maskit, The conformal group of a plane domain, *American Journal of Mathematics*, 90(1968), 718–722.

[MIS] D. Minda and G. Schober, Another elementary approach to the theorems of Landau, Montel, Picard and Schottky, *Complex Variables*, 2(1983), 157–164.

[MUN] J. Munkres, *Elementary Differential Topology*, Princeton University Press, Princeton, NJ, 1963.

[OHS] T. Ohsawa, On complete Kähler domains with C^1 boundary, *Publications of the Research Institute for Mathematical Sciences*, RIMS (Kyoto), 16(1980), 929–940.

[OSS] R. Osserman, A sharp Schwarz inequality on the boundary, *Proceedings of the American Mathematical Society* 128(2000), 3513–3517.

[PRI] I. Privalov, *Randeigenschaften Analytischer Funktionen*, Deutsch Verlag der Wissenschaften, Berlin, 1956.

[ROB] A. Robinson, Metamathematical problems, *Journal of Symbolic Logic* 38(1973), 500–516.

[ROS] B. Rodin and D. Sullivan, The convergence of circle packings to the Riemann mapping, *Journal of Differential Geometry* 26(1987), 349–360.

[RUD1] W. Rudin, *Principles of Mathematical Analysis*, 3rd ed., McGraw-Hill, New York, 1976.

[RUD2] W. Rudin, *Real and Complex Analysis*, McGraw-Hill, New York, 1966.

[RUD3] W. Rudin, *Functional Analysis*, McGraw-Hill, New York, 1973.

[SAZ] R. Saerens and W. Zame, The isometry groups of manifolds and the automorphism groups of domains, *Transactions of the American Mathematical Society* 301(1987), 413–426.

[SPA] E. Spanier, *Algebraic Topology*, Springer-Verlag, New York, 1981.

[STE] K. Stephenson, *Introduction to Circle Packing: The Theory of Discrete Analytic Functions*, Cambridge University Press, New York, 2005.

[STW] E. M. Stein and G. Weiss, *Introduction to Fourier Analysis on Euclidean Spaces*, Princeton University Press, Princeton, NJ, 1971.

[TAO] T. Tao, *Hilbert's Fifth Problem and Related Topics*, American Mathematical Society, Providence, RI, 2014.

[THU] W. P. Thurston, *The Geometry and Topology of 3-Manifolds*, Princeton University Notes.

[WOL] J. Wolf, *Spaces of Constant Curvature*, 5th ed., Publish or Perish Press, Wilmington, DE, 1984.

[ZAL] L. Zalcman, A heuristic principle in complex function theory, *American Mathematical Monthly* 82(1975), 813–817.

Index